Big Data Analytics and Intelligent Systems for Cyber Threat Intelligence

RIVER PUBLISHERS SERIES IN DIGITAL SECURITY AND FORENSICS

The "River Publishers Series in Security and Digital Forensics" is a series of comprehensive academic and professional books which focus on the theory and applications of Cyber Security, including Data Security, Mobile and Network Security, Cryptography and Digital Forensics. Topics in Prevention and Threat Management are also included in the scope of the book series, as are general business Standards in this domain.

Books published in the series include research monographs, edited volumes, handbooks and textbooks. The books provide professionals, researchers, educators, and advanced students in the field with an invaluable insight into the latest research and developments.

Topics covered in the series include-

- Blockchain for secure transactions
- Cryptography
- Cyber Security
- Data and App Security
- Digital Forensics
- Hardware Security
- IoT Security
- Mobile Security
- Network Security
- Privacy
- Software Security
- Standardization
- Threat Management

For a list of other books in this series, visit www.riverpublishers.com

Big Data Analytics and Intelligent Systems for Cyber Threat Intelligence

Editors

Yassine Maleh

University Sultan Moulay Slimane, Morocco

Mamoun Alazab

Charles Darwin University, Australia

Loai Tawalbeh

Texas A&M University-San Antonio, USA

Imed Romdhani

Edinburgh Napier University, UK

LONDON AND NEW YORK

Published 2022 by River Publishers
River Publishers
Alsbjergvej 10, 9260 Gistrup, Denmark
www.riverpublishers.com

Distributed exclusively by Routledge
4 Park Square, Milton Park, Abingdon, Oxon OX14 4RN
605 Third Avenue, New York, NY 10017, USA

Big Data Analytics and Intelligent Systems for Cyber Threat Intelligence / by Yassine Maleh, Mamoun Alazab, Loai Tawalbeh, Imed Romdhani.

Routledge is an imprint of the Taylor & Francis Group, an informa business

ISBN 978-87-7022-778-0 (print)
ISBN 978-10-0084-669-0 (online)
ISBN 978-1-003-37338-4 (ebook master)

Contents

Preface xiii

List of Figures xv

List of Tables xix

List of Contributors xxiii

List of Abbreviations xxvii

Introduction 1
Yassine Maleh, Mamoun Alazab, Loai Tawalbeh, and Imed Romdhani

1 Cyber Threat Intelligence Model: An Evaluation of Taxonomies and Sharing Platforms 3
Hassan Jalil Hadi, Muhammad Adeen Riaz, Zaheer Abbas, and Khaleeq Un Nisa

1.1 Introduction 4
1.2 Related Work 5
1.2.1 Limitations of Existing Techniques 9
1.3 Evaluation Criteria 9
1.3.1 Deployment Setup 10
1.3.1.1 Hardware configurations 10
1.3.1.2 Operating system 10
1.4 Taxonomy of Information Security Data Sources 12
1.4.1 Classification Taxonomy 13
1.4.2 Source Type 13
1.4.3 Information Type 13
1.4.4 Integrability 14
1.5 Trust and Anonymity in Threat Intelligence Platforms 20

1.6 Time (Speed) in Threat Intelligence Platforms (TAXII) . . . 22
1.7 Receiving Time in Threat Intelligence Platforms (TAXII) . . 26
1.8 Conclusion 29
References 29

2 Evaluation of Open-source Web Application Firewalls for Cyber Threat Intelligence **35**
Oumaima Chakir, Yassine Sadqi, and Yassine Maleh
2.1 Introduction 36
2.2 Open-source Web Application Firewalls 38
2.2.1 ModSecurity 38
2.2.2 AQTRONIX Webknight 39
2.3 Research Methodology 40
2.3.1 Implementation of ModSecurity and AQTRONIX Webknight 40
2.3.2 Dataset Description 41
2.3.2.1 Payload All The Thing 41
2.3.3 Experiment Environment 42
2.3.4 Evaluation Metrics 43
2.4 Results and Discussion 43
2.4.1 Results 43
2.4.2 Discussion 45
2.5 Recommendations 46
2.6 Conclusion 46
References 47

3 Comprehensive Survey of Location Privacy and Proposed Effective Approach to Protecting the Privacy of LBS Users **49**
Ahmed Aloui, Samir Bourekkache, Okba Kazar, and Ezedin Barka
3.1 Introduction 49
3.2 Models of Privacy Attack 51
3.2.1 Continuous Location Attack 51
3.2.1.1 Query tracking attack 51
3.2.1.2 Attacks of trajectory 51
3.2.1.3 Identity correspondence 51
3.2.1.4 Location tracking attack 52
3.2.1.5 Attack of maximum movement 52
3.2.2 Context Linking Attack 53
3.2.2.1 Attack of personal context linking 55

3.2.2.2 Attack of observation 53
3.2.2.3 Attack of probability distribution 53
3.3 Mechanisms of Privacy Protection 53
3.3.1 Cloaking 54
3.3.2 Cryptography 54
3.3.3 Obfuscation 54
3.3.4 Dummies 55
3.3.5 Mix-zones 55
3.4 Comparison between Privacy Protection Mechanisms 56
3.5 Types of Environment 57
3.6 Principles of Our Contributions 58
3.7 Our Contribution in Euclidean Space ES 59
3.7.1 Method of Selection of Hiding Candidate Set in ES . 60
3.7.2 Method of Creating Qualified Hiding Region 61
3.7.3 Operation of Our Approach 61
3.7.4 Hiding Principle of Our Approach 62
3.7.5 Generate Dummies (Dummy Queries) 63
3.8 Experimentation 64
3.9 Comparison with Related Works 65
3.10 Conclusion 67
References 67

4 Analysis of Encrypted Network Traffic using Machine Learning Models **71**
Aradhita Bhandari, Aswani Kumar Cherukuri, and Sumaiya Thaseen Ikram
4.1 Introduction 72
4.2 Literature Review 73
4.3 Background 74
4.3.1 Supervised Learning 74
4.3.1.1 AdaBoost 74
4.3.1.2 Random forest 75
4.3.2 Unsupervised Learning 75
4.3.2.1 K-Means clustering 75
4.3.3 Semi-Supervised Learning 75
4.3.3.1 Label propagation 75
4.4 Experimental Analysis 76
4.4.1 Dataset 76
4.4.2 Feature Analysis 76

4.4.3 Pre-Processing 78
4.4.4 Model Results 78
4.4.4.1 K-Means clustering 79
4.4.4.2 Metrics 80
4.4.4.3 AdaBoost 80
4.4.4.4 Random forest 80
4.4.4.5 Semi-Supervised label propagation 81
4.5 Discussion and Future Work 83
4.6 Conclusion 84
References 85

5 Comparative Analysis of Android Application Dissection and Analysis Tools for Identifying Malware Attributes 87
Swapna Augustine Nikale and Seema Purohit
5.1 Introduction 88
5.2 Related Works and Present Contributions 88
5.3 Background and Basic Concepts of Android Ecosystem . . . 89
5.3.1 Android Operating System Architecture 89
5.3.2 Android Application Fundamentals 91
5.4 Android Application Malware Attributes and its Dissection Process 92
5.4.1 Android Application Malware Attributes 92
5.4.2 Android Application Malware Dissection 94
5.5 Android Application Dissection and Malware Analysis Tools 96
5.6 Conclusion and Future Work 100
References 101

6 Classifying Android PendingIntent Security using Machine Learning Algorithms 105
Pradeep Kumar D. S. and Geetha S.
6.1 Introduction 106
6.2 Threat Model 106
6.2.1 Observations 108
6.2.2 Our Contributions 109
6.3 Data Collection and Pre-processing 109
6.3.1 Dataset Discussion 109
6.3.2 Dataset 113
6.3.3 Random Oversampling and Outlier Pre-processing . 113

6.3.4 Correlation Calculation 114
6.4 Identification of Best Machine Learning Model 117
6.4.1 Confusion Matrix 117
6.4.2 Accuracy 117
6.4.3 Precision 118
6.4.4 Recall 118
6.4.5 F1Score 118
6.4.6 AUC-ROC 118
6.5 Discussion 119
6.6 Related Work 122
6.6.1 Limitations and Future Work 123
6.7 Conclusion 123
References 123

7 Machine Learning and Blockchain Integration for Security Applications 129

Aradhita Bhandari, Aswani Kumar Cherukuri, and Firuz Kamalov

7.1 Introduction 130
7.2 Methodology 131
7.3 Background 132
7.4 Blockchain Technology 134
7.4.1 Introduction to Blockchain Technology 134
7.4.2 Applications of Blockchain Technology 136
7.4.2.1 Software-defined network (SDN) specific solutions 137
7.4.2.2 Internet-specific solutions 138
7.4.2.3 IoT-specific solutions 139
7.4.2.4 Cloud storage solutions 139
7.4.3 Smart Contracts 140
7.4.3.1 Blockchain-based smart contracts 140
7.4.3.2 Applications 141
7.4.3.2.1 Internet of Things 141
7.4.3.2.2 Distributed system security . . . 141
7.4.3.3 Finance 142
7.4.3.4 Data Privacy and Reliability 142
7.4.4 Shortcomings of Blockchain Solutions in Cybersecurity 142
7.5 Machine Learning Techniques 144
7.5.1 Introduction 144

7.5.2 Applications in Cybersecurity 144
7.5.2.1 Intrusion detection systems 145
7.5.2.2 Spam detection 146
7.5.2.3 Malware detection 146
7.5.2.4 Phishing detection 147
7.5.3 Shortcomings 147
7.6 Integration of Machine Learning Blockchain Technology . . 147
7.6.1 Blockchain to Improve Machine Learning 148
7.6.2 Machine Learning to Improve Blockchain Solutions 152
7.6.2.1 Machine learning applications in smart contracts 158
7.7 Future Work 159
7.8 Conclusion 161
References 162

8 Cyberthreat Real-time Detection Based on an Intelligent Hybrid Network Intrusion Detection System 175
Said Ouiazzane, Malika Addou, and Fatimazahra Barramou
8.1 Introduction 176
8.2 Related Works 178
8.3 The Proposed Approach 179
8.3.1 Overview of the Overall Architecture of the Previously Proposed System 179
8.3.2 System Components and Its Operating Principle . . 181
8.3.3 Limitations and Points of Improvement of the Old NIDS Model 182
8.3.4 The Proposed Model Architecture 183
8.3.5 Components of the Proposed New Model 184
8.3.6 Operating Principle of the Proposed New Model . . 184
8.4 Experimentation and Results 186
8.4.1 Modeling the Network Baseline 186
8.4.2 Training Dataset – CICIDS2017 188
8.4.3 Classification with the Decision Tree Algorithm . . . 189
8.4.4 Discussion 191
8.5 Conclusion 191
References 192

9 Intelligent Malware Detection and Classification using Boosted Tree Learning Paradigm **195**
S. Abijah Roseline and S. Geetha
9.1 Introduction 196
9.2 Literature Survey 198
9.3 The Proposed Methodology 199
9.3.1 The Rationale for the Choice of Boosting Classifier . 199
9.3.2 Overview 200
9.3.3 Classifiers used for Evaluation 200
9.3.3.1 Decision Tree (DT) 200
9.3.3.2 Random Forest (RF) 201
9.3.3.3 Extra Trees Classifier (ET) 201
9.3.3.4 XGBoost 201
9.3.3.5 Stacked Ensembles 201
9.4 Experimental Results 201
9.4.1 Datasets 201
9.4.1.1 Features of ClaMP Malware Dataset . . . 202
9.4.1.2 Features of BIG2015 Malware Dataset . . 203
9.5 Results and Discussion 205
9.6 Conclusion 208
References 208

10 Malware and Ransomware Classification, Detection, and Prevention using Artificial Intelligence (AI) Techniques **211**
Md Jobair Hossain Faruk, Hossain Shahriar, Mohammad Masum, and Khairul Alom
10.1 Introduction 212
10.2 Malware And Ransomware 214
10.3 Artificial Intelligence 215
10.4 Related Work 216
10.5 Malware Detection Using AI 220
10.6 Ransomware Detection 223
10.6.1 Methodology 223
10.6.2 Experiments and Result 223
10.7 Conclusion 227
References 228

11 Detecting High-quality GAN-generated Face Images using Neural Networks **235**
Ehsan Nowroozi and Yassine Mekdad
11.1 Introduction 236
11.1.1 Organization 237
11.2 State of the Art 237
11.3 Cross Co-occurrences Feature Computation 238
11.4 Evaluation Methodology 240
11.4.1 Datasets 240
11.4.2 Network Architecture 242
11.4.3 Resilience Analysis 243
11.5 Experimental Results 244
11.5.1 Experimental Settings 244
11.5.2 Performance and Robustness of the Detector 245
11.5.3 Performance and Robustness of JPEG-Aware Cross-Co-Net 247
11.6 Conclusion and Future Works 250
References 251

12 Fault Tolerance of Network Routers using Machine Learning Techniques **253**
Harinahalli Lokesh Gururaj, Francesco Flammini, Beekanahalli Harish Swathi, Nandini Nagaraj, and Sunil Kumar Byalaru Ramesh
12.1 Introduction 254
12.2 Related Work 255
12.2.1 Comparative Analysis of Existing Methodologies . . 258
12.3 System Architecture 258
12.3.1 Support Vector Machine (SVM) 260
12.3.2 K-Nearest Neighbor (KNN) 262
12.4 Result Analysis 265
12.5 Conclusion 271
References 271

Index **275**

About the Editors **277**

Preface

The cybersecurity community is slowly leveraging machine learning (ML) to combat ever-evolving threats. One of the biggest drivers for the successful adoption of these models is how well domain experts and users can understand and trust their functionality. Most models are perceived as a black box despite the growing popularity of machine learning models in cybersecurity applications (e.g., an intrusion detection system (IDS)). As these black-box models are employed to make meaningful predictions, the stakeholders' demand for transparency and explainability increases. Explanations supporting the output of ML models are crucial in cybersecurity, where experts require far more information from the model than a simple binary output for their analysis.

In addition, most of the detected intrusions provide a limited set of attributes on a single phase of an attack. Accurate and timely knowledge of all stages of an intrusion would allow us to support our cyber-detection and prevention capabilities, enhance our information on cyber threats, and facilitate the immediate sharing of information on threats, as we share several elements. Big data analytics is an umbrella term for multidisciplinary methods of data analytics that employ advanced mathematical and statistical techniques to analyze large datasets. While data science is a powerful tool to enhance performance, it is only as strong as the data used to develop the solutions. Big data analytics is the intelligence brought to data to transform data into actionable insights for industry and society. Big data analytics benefits organizations in many ways, driving competitiveness and innovation.

On the other hand, big data might contain sensitive and private information that might be at risk of exposure during the analysis process. Big datasets harness information from multiple sources such as databases, data warehouses, and log and event files. Also, data can be obtained from security controls such as intrusion prevention systems and user-generated data from emails and social media posts. So, it is crucial to protect the data of individuals and organizations. The book is expected to address the above issues and present new research in cyber-threat hunting, information on cyber threats, and big data analysis. Therefore, cyber-threat protection of computer systems is one of the most important cybersecurity tasks for single users and businesses, since even a single attack can result in compromised data and sufficient losses. Massive losses and frequent attacks dictate the need for accurate and timely detection methods. Current static and dynamic methods do not provide efficient detection, especially when dealing with zero-day attacks. For this reason, big data analytics and machine-intelligence-based techniques can be used.

Big Data Analytics and Intelligent Systems for Cyber Threat Intelligence comprises many state-of-the-art contributions from scientists and practitioners working in big data intelligence for cybersecurity. It aspires to provide a relevant reference for students, researchers, engineers, and professionals working in this area or those interested in grasping its diverse facets and exploring the latest advances in intelligent systems and data analytics for cyber-threat prevention and detection. More specifically, the book contains 12 chapters. Accepted chapters describe and analyze various applications of big data analytics and *intelligent systems* for cybersecurity applications, such as cyber-threat intelligence, intrusions detection, malware analysis, and Blockchain.

We want to take this opportunity to express our thanks to the contributors of this volume and the editorial board for their tremendous efforts by reviewing and providing interesting feedback. The editors would like to thank Rajeev Prasad, Junko Nakajima, and Nicki Dennis from River Publisher for the editorial assistance and support to produce this important scientific work. Without this collective effort, this book would not have been possible to be completed.

Khouribga, Morocco — *Prof. Yassine Maleh*

Darwin, Australia — *Prof. Mamoun Alazab*

San Jose, USA — *Prof. Loai Tawalbeh*

Edinburgh, UK — *Prof. Imed Romdhani*

List of Figures

Figure 1.1	Deployment diagram	11
Figure 1.2	Sources type	20
Figure 1.3	Taxonomy for information security data source	21
Figure 1.4	Trust in threat intelligence platforms	24
Figure 1.5	Anonymity in threat intelligence platforms	25
Figure 1.6	Time receiving (LAN) in threat intelligence platforms (TAXII)	27
Figure 1.7	Time receiving (Wi-Fi) in threat intelligence platforms (TAXII)	27
Figure 1.8	Time sending (LAN) in threat intelligence platforms (TAXII)	28
Figure 1.9	Time sending (Wi-Fi) in threat intelligence platforms (TAXII)	28
Figure 2.1	The five phases of ModSecurity analysis	39
Figure 2.2	Attack scenario	41
Figure 3.1	Attack of maximum movement	52
Figure 3.2	Tradeoff between privacy and performance	56
Figure 3.3	Difference between Euclidean space and road network space	57
Figure 3.4	Graphe $G(V, E)$	59
Figure 3.5	Clique $C = (S, A)$	60
Figure 3.6	Operation of our approach	61
Figure 3.7	Hiding principle	62
Figure 3.8	Generate Dummies	63
Figure 3.9	Guarantee of privacy	64
Figure 3.10	Quality of service QoS	65
Figure 3.11	Performance	66
Figure 3.12	Comparison with related works	66
Figure 4.1	Visualization of clustering	79
Figure 5.1	Android operating system architecture	90
Figure 5.2	Android application package	93

Figure 5.3 Decompressing APK 94
Figure 5.4 Decompiling .dex/.java files 95
Figure 5.5 Other APK components 95
Figure 5.6 Unzip 96
Figure 5.7 APKTool 96
Figure 5.8 Jadx 97
Figure 5.9 JD-GUI 97
Figure 5.10 Smali/Baksmali 98
Figure 5.11 Dex2Jar 98
Figure 5.12 Enjarify 98
Figure 5.13 Dedexer 98
Figure 5.14 BytecodeViewer 99
Figure 5.15 Android SDK 99
Figure 5.16 Androguard 100
Figure 6.1 PI attack classification 107
Figure 6.2 Data extraction and pre-processing 110
Figure 6.3 Empirical study with CICMalDroid-2020, Drebin, and Google Play Store datasets 110
Figure 6.4 Data extraction architecture 111
Figure 6.5 Percentage of unsafe PI (low value is better) and protected broadcast (high value is better) 112
Figure 6.6 Total broadcast vulnerabilities and PI vulnerabilities (low value is better) 112
Figure 6.7 Dataset snippet (displaying 10 examples) 113
Figure 6.8 Correlation between attributes 115
Figure 6.9 Selecting the best ML model 117
Figure 6.10 AUC-ROC curve (before RandomSampling) 120
Figure 6.11 AUC-ROC curve (after RandomSampling) 121
Figure 7.1 Methodology for the literature review 132
Figure 7.2 Applications of Blockchain in cybersecurity 137
Figure 7.3 Applications of machine learning in cybersecurity . 145
Figure 8.1 Overview of the previous proposed NIDS 180
Figure 8.2 The detection workflow of the previous proposed NIDS 181
Figure 8.3 The new proposed model of hybrid NIDS 183
Figure 8.4 The operating principle of the new proposed model of NIDS 185
Figure 8.5 The followed approach to model the network baseline 187

Figure 9.1 Malware count statistics for windows and android platforms from 2017 to 2021 197
Figure 9.2 Design of the proposed MDS 200
Figure 9.3 Confusion matrix of the proposed MDS for ClaMP malware dataset 206
Figure 9.4 Confusion matrix of the proposed MDS for BIG2015 malware dataset 207
Figure 10.1 Illustration of malware classification 214
Figure 10.2 Ransomware timeline and trends 215
Figure 10.3 Types and uses of artificial intelligence 216
Figure 10.4 Classification of malware detection techniques . . . 220
Figure 10.5 Flowchart of AI-based unknown malware detection techniques 221
Figure 10.6 A high-level view of the VMI-base 222
Figure 10.7 Framework to detect ransomware 224
Figure 10.8 Distribution of the dataset (left) and number of features with varying variance thresholds (right) . . 224
Figure 10.9 ROC curve for decision tree classifier (left) and random forest classifier (right) 227
Figure 10.10 Naïve Bayes classifier (left) and neural network classifier 227
Figure 11.1 The considered CNN-based detector scheme for high-quality GAN-generated face images 240
Figure 11.2 Illustrations of natural FFHQ and StyleGAN2-generated images that are hardly distinguishable . . 241
Figure 11.3 Illustrations of authentic FFHQ and StyleGAN2 generated face images with their printed-scanned versions 242
Figure 11.4 Pipeline of the proposed network architecture . . . 243
Figure 11.5 Visual representation of the Co-occurrence matrices for real image with different channel combinations 246
Figure 11.6 Visual representation of the co-occurrence matrices for GAN image StyleGAN2 with different channel combinations 247
Figure 12.1 Relation diagram of fault error and failure 255
Figure 12.2 Architecture diagram 259
Figure 12.3 Steps performed in fault tolerance detection 260
Figure 12.4 Example of SVM 261

Figure 12.5 Processing of KNN algorithm 264
Figure 12.6 Time cost with an accuracy of hybrid kernel 265
Figure 12.7 Radii with distance calculation using SVM for fault tolerance 266
Figure 12.8 Time cost and accuracy linear kernel 266
Figure 12.9 Fault tolerance of heat map generation using a kernel 267
Figure 12.10 Time cost and accuracy of the single kernel 267
Figure 12.11 Single throughput 268
Figure 12.12 AUC calculation of nodes 268
Figure 12.13 Node failure identification using KNN 269
Figure 12.14 Node failure identification using SVM 270
Figure 12.15 Comparison between SVM and KNN 270

List of Tables

Table 1.1 Hardware specifications 10
Table 1.2 OS specification 10
Table 1.3 Types of sources 17
Table 1.4 Trust and anonymity in threat intelligence platforms . 23
Table 2.1 Assessment of ModSecurity and AQTRONIX Webknight 40
Table 2.2 Number of queries in each payload for each attack . . 41
Table 2.3 CSIC HTTP 2010 Dataset Size 42
Table 2.4 Tools used in this work 42
Table 2.5 SQLI, XSS, and XXE attacks detection using AQTRONIX Webknight 44
Table 2.6 SQLI attacks detection using ModSecurity WAF . . . 44
Table 2.7 XXE attacks detection using ModSecurity WAF . . . 44
Table 2.8 Reflected XSS attack detection using ModSecurity WAF 44
Table 2.9 Performance evaluation of ModSecurity based on CSIC HTTP 2010 dataset 45
Table 2.10 Performance evaluation of AQTRONIX Webknight based on CSIC HTTP 2010 dataset 45
Table 3.1 Comparison of different mechanisms for preserving location privacy 56
Table 4.1 The description of labels in the *attack_cat* column . . 77
Table 4.2 Features extracted after Pearson's correlation 78
Table 4.3 Skewness of the selected features 79
Table 4.4 Confusion matrix for AdaBoost 81
Table 4.5 Confusion matrix of random forest 81
Table 4.6 Confusion matrix for first layer model for label propagation 82
Table 4.7 Confusion matrix for second layer model for label propagation 82

Table 4.8 Comparing the supervised learning techniques and the proposed two-layer label propagation model 83
Table 5.1 Comparative analysis of Android APK dissection and malware attribute analysis tools 100
Table 6.1 Imbalanced data 114
Table 6.2 VIF and tolerance between attributes 116
Table 6.3 The confusion matrix 118
Table 6.4 The outcomes of the algorithms (before RandomSampling)(type of averaging performed on the data = "micro") 120
Table 6.5 The outcomes of the algorithms (after RandomSampling)(type of averaging performed on the data = "micro") 121
Table 7.1 Overview of the applications in which Blockchain was used to improve machine learning technologies 149
Table 7.2 Overview of applications in which machine learning techniques have been used to improve security in Blockchain models, where the last two rows focus primarily on smart contracts 153
Table 8.1 Obtained statistics after applying decision tree to the optimized CICIDS2017 190
Table 8.2 Confusion matrix after applying decision tree to the optimized CICIDS2017 190
Table 9.1 Dataset details of BIG2015 malware dataset 202
Table 9.2 Some important PE malware features in the BIG2015 malware dataset 204
Table 9.3 Comparison of the proposed MDS with other tree-based MDS for the two PE malware datasets . . . 206
Table 10.1 Experimental results analysis of different classifiers . 227
Table 11.1 Accuracies of Cros-Co-Net and Co-Net for unaware and JPEG-aware scenarios over StyleGAN2 and VIPPrint datasets 245
Table 11.2 Robustness performances in the presence of post-processing for StyleGAN2 and VIPPrint dataset . . . 248
Table 11.3 Accuracies of the JPEG-aware Cross-Co-Net for matched and mismatched quality factors for StyleGAN2 and VIPPrint dataset 249
Table 11.4 JPEG-aware Cross-Co-Net robustness performance with the post-processing operators (StyleGAN2). . . 249

Table 11.5 JPEG-aware Cross-Co-Net robustness performance with the post-processing operators (VIPPrint). 250
Table 12.1 Comparative analysis of other methodologies 259
Table 12.2 Experimental setup 265

List of Contributors

Abbas, Z., *Department of Cyber Security and Data Science, Riphah International University, Islamabad, Pakistan; E-mail: zhr670@outlook.com*

Addou, M., *ASYR Team, Laboratory of Systems Engineering (LaGeS), Hassania School of Public Works (EHTP), Casablanca, Morocco; E-mail: malika.addou@gmail.com*

Alom, K., *Northern University Bangladesh, Bangladesh*

Aloui, A., *LINFI Laboratory, Biskra University, Algeria College of Information Technology, UAE; E-mail: a.aloui@univ-biskra.dz*

Barka, E., *LINFI Laboratory, Biskra University, Algeria College of Information Technology, UAE; E-mail: ebarka@uaeu.ac.ae*

Barramou, F., *ASYR Team, Laboratory of Systems Engineering (LaGeS), Hassania School of Public Works (EHTP), Casablanca, Morocco; E-mail: f.barramou@gmail.com*

Bhandari, A., *School of Information Technology and Engineering, Vellore Institute of Technology, India; E-mail: aradhita.bhandari2018@vitstudent.ac.in*

Bourekkache, S., *LINFI Laboratory, Biskra University, Algeria College of Information Technology, UAE; E-mail: s.bourekkache@univ-biskra.dz*

Chakir, O., *Laboratory LIMATI, FPBM, USMS University, Morocco; E-mail: chakir.oumaima@hotmail.com*

Cherukuri, A.K., *School of Information Technology and Engineering, Vellore Institute of Technology, India; E-mail: cherukuri@acm.org*

Faruk, M.J.H., *Kennesaw State University, USA; E-mail: mhossa21@students.kennesaw.edu*

Flammini, F., *University of Applied Sciences and Arts of Southern Switzerland (CH), Switzerland; E-mail: francesco.flammini@supsi.ch*

Geetha, S., *School of Computer Science and Engineering, VIT Chennai, India; E-mail: geetha.s@vit.ac.in*

Gururaj, H.L., *Department of CSE, Vidyavardhaka College of Engineering, India; E-mail: gururaj1711@vvce.ac.in*

Hadi, H.J., *Department of Computer Science, Cyber Reconnaissance Combat, Bahria University, Pakistan; E-mail: hassanjalilhadi1142@gmail.com*

Ikram, S.T., *School of Information Technology and Engineering, Vellore Institute of Technology, India*

Kamalov, F., *School of Electrical Engineering, Canadian University Dubai, UAE; E-mail: firuz@cud.ac.ae*

Kazar, O., *LINFI Laboratory, Biskra University, Algeria College of Information Technology, UAE; E-mail: o.kazar@univ-biskra.dz*

Maleh, Y., *Laboratory LaSTI, ENSAK, USMS University, Morocco; E-mail: yassine.maleh@ieee.org*

Masum, M., *Kennesaw State University, USA; E-mail: hshahria@kennesaw.edu*

Mekdad, Y., *Cyber-Physical Systems Security Lab, Department of Electrical and Computer Engineering, Florida International University, USA; E-mail: ymekdad@fiu.edu*

Nagaraj, N., *Department of CSE, Vidyavardhaka College of Engineering, India; E-mail: nandiningaraj963@gmail.com*

Nikale, S.A., *Department of Computer Science, University of Mumbai, India; E-mail: swapna.nikale@gmail.com*

Nowroozi, E., *Faculty of Engineering and Natural Sciences (FENS), Center of Excellence in Data Analytics (VERIM), Sabanci University, Turkey; E-mail: ehsan.nowroozi65@gmail.com*

Ouiazzane, S., *ASYR Team, Laboratory of Systems Engineering (LaGeS), Hassania School of Public Works (EHTP), Casablanca, Morocco; E-mail: sa.ouiazzane@gmail.com*

Pradeep Kumar, D.S., *School of Computer Science and Engineering, VIT Chennai, India; E-mail: ds.pradeepkumar@yahoo.com*

Purohit, S., *Department of Computer Science, University of Mumbai, India; E-mail: supurohit@gmail.com*

Ramesh, S.K.B., *Department of CSE, Vidyavardhaka College of Engineering, India; E-mail: sunilkumar.br@vvce.ac.in*

Riaz, M.A., *Department of Information Technologies, KFUEIT, RYK, Pakistan*

Roseline, S.A., *School of Computer Science and Engineering, Vellore Institute of Technology - Chennai Campus, India;*
E-mail: abijahroseline.s2017@vitstudent.ac.in

Sadqi, Y., *Laboratory LIMATI, FPBM, USMS University, Morocco;*
E-mail: yassine.sadqi@gmail.com

Shahriar, H., *Kennesaw State University, USA;*
E-mail: hshahria1@kennesaw.edu

Swathi, B.H., *Department of CSE, Vidyavardhaka College of Engineering, India*

List of Abbreviations

ADNIDS	Anomaly detection NIDS
AHE	Adaptive histogram equalization
AI	Artificial intelligence
AOSP	Android open source projects
AOT	Ahead-of-time
API	Application programming interface
APK	Android application package
APT	Advanced persistent threats
ART	Android runtime
AUC	Area under the curve
BAD	Blockchain anomaly detection
BHO	Browser helper object
CAPEC	Common attack pattern enumeration and classification
CIF	Collective intelligence framework
CNN	Convolutional neural network
CRITs	Collaborative research into threats
CRS	Core Rule Se
CSIRTs	Computer security incident response teams
CTI	Cyber threat intelligence
CybOX	Cyber Observable EXpression
D3NS	Distributed decentralized domain name system
DANN	Deep autoencoder neural networks
DBIR	Data breach investigations report
DDoS	Distributed denial-of-service
DL	Deep Learning
DLLs	Dynamic link libraries
DNN	Deep neural network
DNS	Domain name system
DoH	DNS over HTTPS
DoS	Denial of service

DPGAN	Differentially private generative adversarial networks DR demand response
DR	Demand response
DT	Decision tree
ET	Extra trees
FA	Filtering agent
FEA	Feature extraction agent
FFHQ	Flicker-Faces-HQ
FN	False negative
FP	False positive
FPPDL	Fair and privacy preserving deep learning
FPR	False positive rate
FTDR	Fault-tolerant deflection routing algorithm
GAN	Generative adversarial networks
HAL	Hardware abstraction layer
HTTP	Hypertext transfer protocol
HTTPS	Hypertext transfer protocol secure
ICC	Inter-component communication
IDMEF	Intrusion detection message exchange format
IDS	Intrusion detection system
IOC	Indicator of compromise
IODEF	Incident object description exchange format
IoT	Internet of Things
IPC	Inter-process communication
IPFS	Interplanetary file system
ISO	International standards organization
JD-GUI	Java decompiler – graphical user interface
KNN	K-nearest neighbor
LaaS	Law as a service
LBS	Location-based services
LFIL	Link fault identification and localization
LR	Logistic regression
LSTM	Long-short term memory
MAC	Mandatory access control
MAEC	Malware attribute enumeration and characterization
MANs	Metropolitan area networks
MBR	Minimum bounding rectangle
MISP	Malware information sharing platform

ML	Machine learning
MLP	Multilayer perceptron
MSM	Making security measurable
NB	Naïve Bayes
NIDS	Network intrusion detection system
NLP	Natural language processing
NN	Neural Network
NOC	Network-on-chip
OASIS CTI TC	OASIS cyber threat intelligence technical committee
OpenC2	Open command and control
OpenIOC	Incident of compromise
OWL	Web ontology language
P2P	Peer-to-peer
PE	Portable executable
PFM	Probability-based factor model
PI	PendingIntent
PII	Personally identifier information
PIR	Personal information retrieval
PKI	Public key infrastructure
PL	Paranoia level
PMF	Probabilistic matrix factorization
PoAI	Proof-of-artificial intelligence
QoS	Quality of services
QP	Quadratic programming
ReLu	Rectified linear unit
RF	Random forest
ROC	Receiver operating characteristics
SA	Sniffer agent
SDN	Software-defined network
SELinux	Security-enhanced linux
SNIDS	Signature detection-based NIDS
SOC	System-on-chip
SOMs	Self-organizing maps
SQLI	SQL injection
STIX	Structured threat information eXpression
SVM	Support vector machine
TAXII	Trusted Automated Indicator Information Exchange

TAXII	Trusted Automated eXchange of Information
TE	Traffic engineering
TISPs	Threat intelligence sharing platforms TIP
TN	True negative
TP	True positive
TPR	True positive rate
TTP	Trusted third party
UAV	Unmanned aerial vehicle
UI	User interface
UIR	Unauthorized intent receipt
UPSIDE	Universal pattern for structured incident exchange
VEPM	Version evolution progress model
VERIS	Vocabulary for event recording and incident sharing
WAF	Web application firewall
WANs	Wide area networks
WI	WrappingIntent
XGB	Extreme gradient boosting
XSS	Cross-site scripting
XXE	XML external entities

Introduction

Yassine Maleh, Mamoun Alazab, Loai Tawalbeh, and Imed Romdhani

As cyberattacks against critical infrastructure increase and evolve, automated systems to complement human analysis are needed. Such organizations are so large with so much information and data to sort through to obtain actionable information that it seems impossible to know where to start. Moreover, chasing the breaches is like looking for a needle in a haystack. The intelligence analysis of an attack is traditionally an iterative, mainly manual, process, which involves an unlimited amount of data to determine the sophisticated patterns and behaviors of intruders. In addition, most of the detected intrusions provide a limited set of attributes on a single phase of an attack. Accurate and timely knowledge of all stages of an intrusion would allow us to support our cyber-detection and prevention capabilities, enhance our information on cyber threats, and facilitate the immediate sharing of information on threats, as we share several elements.

Big data analytics is an umbrella term for multidisciplinary methods of data analytics that employ advanced mathematical and statistical techniques to analyze large datasets. While data science is a powerful tool to enhance performance, it is only as strong as the data used to develop the solutions. Big data analytics is the intelligence brought to data to transform data into actionable insights for industry and society. Big data analytics benefits organizations in many ways, driving competitiveness and innovation.

The cybersecurity community is slowly leveraging machine learning (ML) to combat ever-evolving threats. One of the biggest drivers for the successful adoption of these models is how well domain experts and users can understand and trust their functionality. Despite the growing popularity of machine learning models in cyber-security applications (e.g., an intrusion detection system (IDS)). As these black-box models make meaningful predictions, the stakeholders' demand for transparency and explainability increases.

Explanations supporting the output of ML models are crucial in cybersecurity, where experts require far more information from the model than a simple binary output for their analysis.

On the other hand, big data might contain sensitive and private information that might be at the risk of exposure during the analysis process. Big datasets harness information from multiple sources such as databases, data warehouses, and log and event files. Also, data can be obtained from security controls such as intrusion prevention systems and user-generated data from emails and social media posts. So, it is crucial to protect the data of individuals and organizations. Therefore, cyber-threat protection of computer systems is one of the most critical cybersecurity tasks for single users and businesses, since even a single attack can result in compromised data and sufficient losses. Massive losses and frequent attacks dictate the need for accurate and timely detection methods. Current static and dynamic methods do not provide efficient detection, especially when dealing with zero-day attacks. The book is expected to address the above issues and present new research in cyber threat hunting, information on cyber threats, and big data analysis.

For this reason, big data analytics and machine-intelligence-based techniques can be used. This book brings together researchers and professionals in the field of big data analytics and intelligent systems for cyber-threat intelligence (CTI) and key data to advance the missions of anticipating, prohibiting, preventing, preparing, and responding to internal security. The wide variety of topics it presents offers readers multiple perspectives on various disciplines related to big data analytics and intelligent systems for cyber-threat intelligence applications.

This book proposes practical solutions for an effective cyber-threat intelligence strategy, taking into account recent technological developments such as the Internet of Things, Blockchain, etc., in order to help decision-makers, managers, cybersecurity professionals, and researchers design a new cybersecurity paradigm, taking into account the new opportunities associated with machine intelligence and big data analytics. It combines theoretical concepts and practical use cases, so that readers of the book (beginners or experts) find both a description of the concepts and context related to cyber-threat intelligence.

Keywords: Big Data Analytics, Machine Learning, Cybersecurity, Cyber-Threat Intelligence, Intelligence Analysis, Intrusion Detection.

1

Cyber Threat Intelligence Model: An Evaluation of Taxonomies and Sharing Platforms

Hassan Jalil Hadi[1,2,3], Muhammad Adeen Riaz[4], Zaheer Abbas[5], and Khaleeq Un Nisa[6]

[1]Department of Computer Science, Bahria University, Pakistan
[2]Cyber Security and Data Science, Bahria University, Pakistan
[3]Cyber Reconnaissance Combat, Bahria University, Pakistan
[4]Department of Information Technologies, KFUEIT, RYK, Pakistan
[5]Department of Riphah International University Islamabad, Pakistan
[6]Department of Cyber Security and Data Science, Riphah International University, Pakistan
E-mail: Hjalil.buic@bahria.edu.pk; m.adeen@kfueit.edu.pk; zhr670@outlook.co; engrkhaleeq54@gmail.com

Abstract

To defend assets, data, and information against state-of-the-art and increasing number of cyber threats, cyber defenders should be one step ahead of these cybercriminals. This phase is possible if and only if the cyber defender gathers enough information about threats, risks, vulnerabilities, attacks, and countermeasures on time or before an incident is going to happen. Cybersecurity staff collects cyber threat information from multiple sources, extending from inter-organizations to publicly available sources, threat intelligence sharing platforms such as mailing lists or expert blogs, etc. Intelligence provides evidence-based knowledge regarding potential or existing threats. The advantages of threat intelligence are the effectiveness of security operations and better efficiency in detection and prevention abilities. Good threat intelligence for cyber domains requires a knowledge base containing threat information and an effective way to present this knowledge. For this purpose, taxonomies, ontologies, and sharing platforms are used. The proposed cyber threat intelligence model enables cybersecurity experts to investigate their capabilities for threat intelligence and comprehend their position against the continuously changing landscape of cyber threats. Moreover, this model is used for analyzing and evaluating numerous existing sharing platform taxonomies, ontologies, and sharing platforms related to cyber threat intelligence. The results indicate a need for an ontology that covers the whole spectrum of CTI in the community of cybersecurity.

Keywords: Cyber threat intelligence, threat intelligence sharing platform, information security data sources, cyberattack attribution.

1.1 Introduction

To prevent these threats, various vulnerability management steps are implemented by organizations, conducting awareness training sessions that are part of their information security program [2], [3]. It is important to highlight that these countermeasures must be improved, leading to trends like collaboration, automation, and consultation of shared data security sources. Various practices and research methodologies are needed to achieve these goals [4], [5]. Public data security information sources are defined as sources of information providing information about the possible threats, risks involved, attacks, vulnerabilities, countermeasures available, and the assets affected (ISO and Std, 2009). Various frameworks, technologies, data formats, and message protocols have been introduced into research and practice, allowing automated and standardized information security exchanges. There has been very little research on public "information security" data sources and the sharing of threats. As a result, several platforms have entered the market with "threat intelligence sharing." Sauerwein *et al.* [6] highlight that information security data can be accessed via these platforms and found on the public "information security" data sources.

This study's main objective is to identify and address gaps by analyzing publicly available threat intelligence sharing platforms. A consequence of this is the unavailability of public data security sources and their dependencies are not systematically or comprehensively reviewed by research and practice. Moreover, a proper scheme of classification is needed for these information security data sources so that it is possible to compare them with the previous ones.

To understand the breadth of the variety of "threat intelligence sharing platforms (TISPs)" standards, sources, challenges, and gaps, an extensive analysis was performed:

- Defined the performance matrix of the TAXII-enabled platforms
- Implementation of TAXII-enabled platforms (open source free to use)
- Performance measurements on the following parameters:
    - Trust
    - Anonymity
    - Time (speed)

Classification scheme for the information security data sources is based on the following key aspects:

- Information type
- Timeliness
- Integrability
- Originality
- Source type
- Trustworthiness

1.2 Related Work

Cyber threat intelligence (CTI) provides comprehensive analysis of organizations, institutes, and enterprises that have conducted an academic and business perspective. Various contributions directly linked to the "information security" data sharing process can be identified. These contributions focus on analyzing different models related to sharing "information security," data sources, sharing of "threat intelligence," and other related platforms. A study has been conducted on analyzing the use cases related to the scalability and interoperability of information security exchange and protocol. Steinberger *et al.* [8] has given a formal analysis of this current information. Furthermore, models for sharing information have also been proposed by Hernandez-Arrieta *et al.* [9] in real-time build on the making security measurable (MSM).

Qamar *et al.* [10] and Matters *et al.* [11] highlight the proactive approach of cyber threat intelligence (CTI) to predict threats and enhance knowledge to future coup challenges of attacks. The research aims to provide a mechanism for threat analytics based on "Web Ontology Language (OWL)" for automated evaluation of large volumes of CTI data. The suggested framework was intended to identify the threat's significance and probability and evaluate the risk to a network. Burger *et al.* [12] also discuss the CTI scientifically. The authors provide a taxonomy model to classify, recognize, and disclose the variations between current CTI techniques in the classification model needed for "cyber threat information" exchange technologies. Tounsi and Rais [13] have given various threat intelligence forms. Their focus was formulating new standards, introducing recent trends, and analyzing technical issues. Likewise, Mavroeidis *et al.* [7] successfully developed a classification scheme for the ontologies and sharing standards. Research and practices have implemented various data formats, frameworks, message protocols, and technologies that allow the automated and standardized exchange of data regarding information security [14].

Menges and Pernul established a processing model to compare different incident reporting formats. Depending on the Universal Pattern for Structured Incident Exchange (UPSIDE) model, they created a set of structural requirements to compare incident reporting formats. They choose STIX 1.0, STIX 2.0, IODEF 1.0, IODEF 2.0, VERIS, and X-ARF models for analysis. Participating researchers have different backgrounds and objectives, indicating their interest in CTI. It is a key element for improvement in cybersecurity mechanisms. Proponents argue that CTI can help organizations:

- To prepare smartly against cyberattacks
- To be aware of the present and future threats
- To efficiently detect existing threats
- To better prepare and understand challenges/attacks
- To prepare accurate and comprehensive defense strategies

CTI must be actionable to meet its full potential. Researchers suggested that CTI is useless if it does not justify its action or decision. It is highlighted that CTI must be relevant. If an organization CTI is not relevant from an institutional or technical perspective, then collecting CTI to that specific organization is useless. Timeliness of CTI data is a significant concern [12]. Attackers develop new techniques and tools to compromise systems and networks; so data relevance is significant. Therefore, consumers need to update their records. It is mentioned that a great variety of external CTI sources are available. External sources have two categories, i.e., commercial and open-source. Governmental and non-governmental sources are another distinction. CTI is based on organizations' objectives and requirements. CTI platforms are available for purchasing and exchanging data in organizations. Different platform solutions are available in the market with a range of functionalities.

CTI is a broad concept without a common understanding or definition. However, Sauerwein *et al.* [6] identified five different platforms focusing on IOC sharing. However, the authors claim that most tools primarily focus on data collection and more or less neglect the intelligence lifecycle. Therefore, provided data cannot count intelligence in its strictest sense. CTI sharing platforms, according to them, provide restricted capacity for assessment and visualization. They lag behind platforms for information sharing and information mining solutions from other fields. In a nutshell, what finds CTI lies in the eye of the perceiver.

The organization ensures quality of CTI to meet organizational requirements and needs before subscribing to a CTI data feed, framework, or participating in a CTI community. It is important to ensure that consumed data should be timely, relevant, and effective for better active CTI [16]. It is important to highlight how, when, and from where the data is sourced. These questions should be in security practitioners' minds before selecting CTI sources for the organization. Validation of manually generated high volumes of complex CTI data becomes challenging. It is noted that the mentioned problem is often redundant, incomplete, or even worse concerning CTI [10]. According to Qamar *et al.*, this data redundancy and non-uniformity make analyzing a sample CTI report more difficult to identify its significance to a specific network [10].

Advanced persistent threats (APT) and the fast-changing cyberattack landscape mutual exchange of relevant cyber threats information. Mostly, exploiting weaknesses exists in multiple systems, products, or networks rather than launching attacks on a single target. Hence, information regarding how, when, and why organizations have been compromised is important. The same attack can be preventable in other organizations based on the provided information. Mutual exchange of information among organizations is a basic pillar of CTI. Cyber threats intelligence should meet specific requirements, i.e., provided information should be relevant, actionable, and time-oriented. Quality issues like redundancy, inaccuracy, and incorrectness hamper and compromise the effectiveness of the CTI framework. In these circumstances, it is inevitable to incorporate standardized methods for producing and sharing CTI data.

Researches and developments on establishing standards and frameworks for sharing CTI data focus on developing a standardized language, structure for documenting cyber threats, and techniques and tools to enable automated and standardized distribution.

The most promising cyber threat intelligence standards on the market today are the "Trusted Automated eXchange of information" (TAXII) and "Structured Threat Information eXpression" (STIX). STIX and TAXII are open community efforts that are constantly developing. The MITRE Corporation initially served on behalf and under the sponsorship of the US Homeland Security Department as the creator and moderator of both the STIX and TAXII communities. US Homeland Security Department, in 2015, moved STIX and TAXII for further development to the "OASIS Cyber Threat Intelligence Technical Committee (OASIS CTI TC)." Today, STIX and TAXII are widely discussed among the CTI communities. Moreover, they have already been incorporated into many CTI platforms and tools, a circumstance that makes them a *de-facto* standard for CTI sharing and its consumption [17], [18]. Many researchers have identified the challenges and needs of threat intelligence sharing platforms [19]–[22].

Furthermore, various standardization efforts to enable the uniform sharing of cybersecurity data, for instance, "Structured Threat Information eXpression" (STIX), "Trusted Automated eXchange of information" (TAXII), "Cyber Observable EXpression" (CybOX), "Incident of compromise" (OpenIOC), or "Incident Description Exchange Format" (IODEF) [23], [24]. For correct standardization of a specific application, Burger *et al.* introduced a framework with the help of which standards can be analyzed and evaluated. Furthermore, such standards form the foundation for the current threat intelligence sharing platforms [12].

Studies related to platforms of threat intelligence sharing are very limited. In [25], the SANS Institute outlines a minor selection of an open-source TIP. They also include the "Collaborative Research into Threats" (CRITs), "Malware Information Sharing Platform" (MISP), "MANTIS Cyber-Intelligence Management Framework," "Collective Intelligence Framework" (CIF), and SoltraEdge, concluding that the information exchanging landscape is still undergrowth. Brown *et al.* described numerous difficulties linked to platforms for sharing threat intelligence, community needs, and expectations [20]. These implications focus solely on a subset of existing platforms for sharing threats without comprehensive modern technology analysis. Some studies related to threat intelligence sharing in organizations, such as software suppliers in governance management software, risk, and compliance software (GRC), the security software market, and cloud providers, can be found.

Wagner *et al.* [26] introduced a generic cyber threat sharing platform that allows the distribution of cyber threat intelligence based on anonymity. They analyzed threat sharing platforms regarding anonymity. Wagner *et al.*

[27] presented a new trust classification to establish a trustworthy threat sharing atmosphere. They analyzed and compared threat intelligence platforms/providers regarding trust functionalities. Abu *et al.* [28] carried out a literature review on current threat intelligence definitions. Their findings have shown that organizations and vendors lack knowledge of what data should be regarded as intelligence risk. Johnson *et al.* [29] delivered recommendations on contributing to threat intelligence sharing. Sauerwein *et al.* [6] performed and compared a systematic survey of various platforms for sharing threat intelligence. Furthermore, in the practice of "threat intelligence" sharing, Sillaber *et al.* [30] acknowledged data quality assays. Lastly, Sauerwein *et al.* [31] introduce the "shadow-threatening intelligence" reports, which are discussed by outlining several information security data sources.

1.2.1 Limitations of Existing Techniques

- Does not provide a comparative analysis of threat intelligence sharing platforms:
  - Deployment complexity
  - Does not mention which platform is easy to use and easy to deploy
  - Does not suggest which platform is best based on given parameters
- Organizations are unable:
  - To comprehend which platform is best based on given parameters
  - To select the most suitable threat intelligence sharing platform to participate in collaborative security.

1.3 Evaluation Criteria

Thorough research provides a comparative analysis of threat intelligence sharing platforms. Following will be the steps:

- Literature review.
- Gather and report all possible threat intelligence sharing sources, frameworks/platforms, standards, and formats.
- Select threat intelligence sharing platforms, TAXII supported, open-source free to use, or closed source trial available.
  - MISP, OpenTAXII, SoltraEdge, Anomaly STAXX, SRA TAXII 2.0 Server, CABBY, Tripwire, FreeTAXII 2.1, OASIS CTI Client, OASIS CTI Server, and Splunk Enterprise 7.2.4.

- Evaluation parameters on which threat intelligence sharing platforms will be analyzed are time (speed), trust, and anonymity, and also taxonomy for the classification of information security data.

After selecting parameters, collect and extract information of threat intelligence sharing platforms on selected parameters.

- Analysis and results
- Compile research

1.3.1 Deployment Setup

1.3.1.1 Hardware configurations

For the deployment and implementation, we used HP ProDesk 400 G5 MT PC having the following specification:

Table 1.1 Hardware specifications

Name	Description
CPU	"Intel (R) Core(TM) i7-8700 CPU @ 3.20 GHz 1 physical processor; 6 cores; 12 threads"
RAM	16138488 KiB
Motherboard	HP83FO
Graphics	Mesa DRI Intel(R) UHD Graphics 630 (Coffeelake 3 × 8 GT2)
Ethernet Card	RealTek RTL-8169 Gigabit Ethernet
Storage	ATA SPCC Solid State Drive 512 GB

1.3.1.2 Operating system

For the deployment and implementation, we used Linux having the following specification:

Table 1.2 OS specification

Name	Description
Kernel	Linux 4.15.0-47-generic (×86-64)
Version	50-Ubuntu SMP Wed Mar 13 10:44:52 UTC 2020
C Library	GNU C Library/(Ubuntu GLIBC 2.27-3ubuntu1) 2.27
Distribution	Linux Mint 19.1 Tessa

In research, a detailed literature review was carried out at first. In the literature review, the authors identify various contributions directly linked to the field of "information security" data sharing processes. Then they elaborate limitations of existing techniques and extract the problem statements. Develop Comparative topic Analysis of Threat Intelligence Sharing

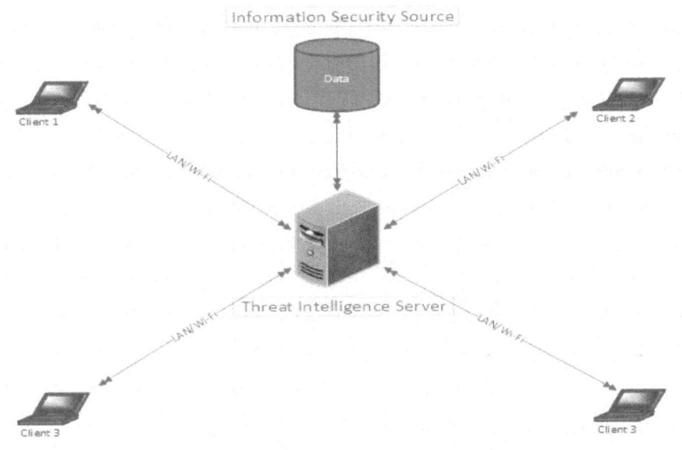

Figure 1.1 Deployment diagram

Platforms. Then find literature on that and extract data from multiple papers. Trust, anonymity, and time (speed) are three parameters compared for analysis. For the parameter time (speed), shortlist threat intelligence sharing platforms, TAXII supported, open-source free to use, or closed source trial available. For trust and anonymity, collect data of 32 threat intelligence sharing platforms from different sources and papers and present it in the form of a table with some explanation.

Successfully able to deploy and implement the following models: Yeti (TAXII 1.0), OpenTAXII (TAXII 1.0), Cabby (TAXII Client 1.0/1.1), CTI TAXII Server 2.0, CTI TAXII Client 2.0, STIX 2.0 Data Generator, STIX Visualizer, FreeTAXII 2.1, SRA TAXII 2.0 Server, Anomaly STAXX, MISP, SoltraEdge, and Splunk. Although that was not an easy task, there was limited help available on deploying these platforms. There were many dependencies after many tires successfully deployed and configured these platforms. Some changes were made in the configurations and integrated CTI TAXII Server 2.0 with STIX 2.0 Data Generator to use that STIX 2.0 data as input and send data from one end to another. FreeTAXII 2.1 SRA TAXII 2.0 Server was using HTTPS over TLS. So, create our self-signed certificate through OpenSSL and integrate it with CTI TAXII Server 2.0, which earlier had that HTTP could not work on HTTPS over TLS. This task helps us in privacy and data integrity between our communicating parties. Anomaly STAXX, MISP, SoltraEdge, and TripWire were available in virtual machines and after some necessary configurations, they came up and running. Splunk deployed and configured after its necessary dependencies. MISP is a good open-source threat intelligence sharing platform that is expandable. It can import many types of formats and export them into many formats. It can easily integrate with other threat intelligence sharing platforms, such as NIDS and LIDS, and log analysis tools such as SIEMs.

Authors integrate CTI TAXII Server 2.0 through its rest API and the authorization header. Now, the second source of input to CTI TAXII Server 2.0. First, MISP data is converted to STIX 2.0 and from API, MISP data is sent through CTI TAXII Server 2.0 to another end. Then, calculate the time of receiving information from platforms. OpenTAXII (TAXII 1.0) uses Cabby (TAXII Client 1.0/1.1) to retrieve data. OpenTAXII has static data STIX 1.0, which loads from the file. CTI TAXII Server 2.0 uses CTI TAXII Client 2.0 for retrieving and sending data. FreeTAXII 2.1 and SRA TAXII 2.0 Server also have static data. They can store more data, but it is not possible to add or send data to the server in the given configurations. Maybe in the coming future, it will. SoltraEdge, Anomaly STAXX, and Splunk show data in web view, and data can be added through the import interface. Anomaly STAXX gets data from its server LIMO or imports data in text format, while SoltraEdge and Splunk support STIX 1.0 data format.

These three are the trial version and have less functionality. Then, calculate the time of sending information to a server or other clients. In this, only CTI TAXII, Server 2.0, and MISP are left, and they can send and receive data. So, first send data to CTI TAXII Server 2.0 with STIX 2.0 Data Generator's STIX 2.0 data and then MISP data to CTI TAXII Server 2.0. After that, send data to MISP and calculate time. It took more time than CTI TAXII Server 2.0 because it contains more attributes. The average amount of data and the internet speed may affect the time of sending. Then, we collected a list of 127 data sources of information security for analysis. To classify these data sources, we created a taxonomy that includes the following features:

- Information type
- Timeliness
- Integrability
- Originality
- Source type
- Trustworthiness

1.4 Taxonomy of Information Security Data Sources

Here, we present the classification that has been developed and the outcomes of the assessment of data sources for "information security."

1.4.1 Classification Taxonomy

Following features are included in the classification scheme of data sources for information security:

- Information type
- Timeliness
- Integrability
- Source type
- Originality
- Trustworthiness

1.4.2 Source Type

Various types of data sources for information security are categorized. These nine types were produced because of the incremental process during classification. We, therefore, distinguish between these 13 types as follows:

- Vulnerability database
- Vendor website (security product)
- Mailing list archive
- Expert security blogs
- Discussion groups
- Professional forums
- Hacktivist groups
- Deep-web
- Darknet sources
- Streaming portals
- News websites
- Social network and others

1.4.3 Information Type

To classify the sort of information supplied from information security data sources, we differentiate these sorts of information from IEC (2014) as they signify the most prevalent types of data in information security:

i. *Asset* : Data about any item or characteristic information of value to a company.
ii. *Vulnerability* : Weakness information of a threat that may be exploited.
iii. *Risk* : Information unfolding possible event consequences for an attack.

iv. *Threat* : Information about the possible cause of an unwanted incident.
v. *Attack* : Data on any illegal effort to gain access to, modify, or terminate an asset.
vi. *Countermeasure* : Data on managerial, administrative, legal control, or technical used to respond to the threat of information security.

A data source could provide more than one sort of data. As a result, it would be possible to have multiple classifications of the information type. A vulnerability database, for example, could deliver information about vulnerabilities and threats.

1.4.4 Integrability

The integrability of information is unavoidable to automatically manage information security processes, as outlined in "IEC/ISO 27005 ISO/IEC (2011)." Our context describes how (automatically) data sources for information security and the information delivered can be combined into the landscape and procedures of an organization's data security tool [33]. The classification scheme gives us the format of accessible information and the interfaces given to determine the integrability of data sources for information security. The taxonomy separates the following two kinds to classify the accessible data format.

i. Structured: Security information in the organized format is accessible. We, therefore, distinguish between these eight types as follows:

a) **Common Attack Pattern Enumeration and Classification (CAPEC):** "It is a comprehensive dictionary and analysis approach to understand the attack pattern that leads to enhance defense by using classification taxonomy of known attacks.
b) **Cyber Observable eXpression (CybOX):** "This language provides a common structure for representing cyber observables across and among the operational areas of enterprise cybersecurity that improves the consistency, efficiency, and interoperability of deployed tools and processes, as well as increases overall situational awareness by enabling the potential for detailed automatable sharing, mapping, detection, and analysis heuristics.
c) **Incident Object Description Exchange Format (IODEF):** "It is a data representation technique that provides a framework for sharing information commonly exchanged by computer security incident response teams (CSIRTs) about computer security incidents.

d) **Malware Attribute Enumeration and Characterization (MAEC):** This pattern aims to create and provide standardized language for sharing structured information based on attributes such as behaviors, artifacts, and attack patterns about malware.
e) **Open Command and Control (OpenC2) Technical Committee:** The OpenC2 TC was chartered to draft lexicons, specifications, documents, or other artifacts to fulfill the needs of cybersecurity command and control in a standardized manner.
h) **Structured Threat Information eXpression (STIX) Language:** "It is a standardized construct language to represent cyber threat information and convey the full range of potential cyber threat information which strives to be fully expressive, flexible, extensible, and automatable. It also provides test mechanisms for embedding tool-specific elements, including Open IOC, Yara, and Snort.
g) **Vocabulary for Event Recording and Incident Sharing (VERIS):** Set metrics designed for common language to describe security incidents in a structured and repeatable manner. It is a response to one of the most critical and persistent challenges in the security industry. It is also highlighted that VERIS collects data from the community to report on breaches in the Verizon Data Breach Investigations Report (DBIR) and publishes this database online at VCDB.org.
h) **Intrusion Detection Message Exchange Format (IDMEF):** "The purpose of IDMEF is to define data formats and exchange procedures for sharing information of interest to intrusion detection and response systems.

ii. Unstructured: Without using a common data representation format, security information is accessible. To classify the interfaces supplied, the taxonomy distinguishes between the following four kinds:

a) **No interfaces:** Source of information security does not deliver any access interface.
b) **APIs:** "Application programming interface" (API) is provided by the "information security" data source to get the provided data.
c) **RSS Feed:** Data source information security offers RSS Feeds for tracking/monitoring the data that has been provided.
[d) **Export:** Data source for information security offers an interface for exporting content as JSON, XML, CSV, HTML, IDS/IPS Rules, PDF, or plain text.

There may be more than one interface available from a source of information. Therefore, several classifications of the provided interfaces would be possible.

iii. Trustworthiness

In areas related to information security, trustworthiness is important when sharing information because wrong information could have an enormous influence on organizational procedures [34]. Consequently, we investigated the (1) traceability, (2) credibility (in [33], the definition of these two terms), and (3) feedback mechanisms. For classifying publishers (1), taxonomy distinguishes among (a) the government, (b) the vendor, (c) the normal user, and (d) the security experts. It is noteworthy that if there is enough background information available on the publisher, the sort of publisher would be categorized as (a), (b), or (d). If not, the publisher would categorize it as (c). In addition, it may be possible that there might be more than one sort of publisher in the information source. Second, if the data is traceable, the taxonomy is classified. So, security information is classified as traceable, if a publisher and publishing date can be found back, depending on the metadata. The information shall otherwise classify as non-traceable. Lastly, we also analyze whether the data is validated. So now, the data is classified as (3) provided by feedback mechanisms, which comprise user ratings and comments concerning the effectiveness of the data supplied.

iv. Originality

For selecting on the originality and creativity of the information (cf., [33]), the taxonomy classifies the origin into two sources:

a) **Original source:** Data derived from the data source for information security (database of national vulnerability [35] who distributes their data).
b) **Secondary source:** Data from other information security data sources is copied or incorporated, for example, from a meta-source that incorporates information from various originals.

v. Timeliness

In information security risk management, timeliness plays an important role because the necessary data (for example, regarding threats) should be given at an appropriate time to take suitable measures (cf., [36] and [37]). For example, if an organization is notified of an emergency risk as timely as possible, counter-actions can be implemented and an attacker can be prevented early.

Our second dimensions, therefore, offer a closer analysis of the performance of sharing as far as time is concerned. The taxonomy differentiates the following two classes that came from the work of White and Zhao [38] to classify the sharing behavior:

a) **Routine:** Information is published regularly at a specific time, for instance, daily, weekly, or monthly reports.
b) **Incident-specific:** Information is published when news is available or new incidents occur.

Table 1.3 Types of sources

Sr. No.	Sources	VU DB	Website vendor	List mailing	Blog	Forum	Portal for streaming	News website	Social net	Others
2	Abuse IP DB									✓
3	Apility.io									✓
4	APT Groups and Operations									✓
5	AutoShun									✓
6	Akaoma									✓
7	AlephSecurity				✓					
8	BoosteroK				✓					
9	Binary Defense IP Banlist									✓
10	BGP Ranking									✓
11	Botnet Tracker									✓
12	BOTVRIJ.EU									✓
13	BruteForce Block									✓
14	Bugtrack									✓
15	C&C Tracker									✓
16	CCSS Forum Malware Certificates					✓				
17	CI Army List									✓
18	Cisco Umbrella									✓
19	Critical Stack Intel					✓				
20	C1fApp					✓				
21	CVE Details	✓								
22	CyberCure free intelligence feeds									✓
23	CyberWarZone							✓		
24	DarkReading							✓		
25	Dshield									✓
26	Disposable Email Domains									✓
27	Debian		✓							
28	DNSTrails				✓					
29	Emerging Threats Firewall Rules									✓

(Continued)

Table 1.3 Continued.

Sr. No.	Sources	VU DB	Website vendor	List mailing	Blog	Forum	Portal for streaming	News website	Social net	Others
30	ENISA									✓
31	ExoneraTor		✓							
32	Exploitalert	✓								
33	ZeuS Tracker	✓								
34	FireHOL IP Lists					✓				
35	FraudGuard		✓							
36	Grey Noise		✓							
37	Hail a TAXII									✓
38	HoneyDB									✓
39	Icewater	✓								
40	Infosec-CERT-PA	✓								
41	I-Blocklist									✓
42	Majestic Million									✓
43	Malc0de DNS Sinkhole	✓								
44	MalShare.com					✓				
45	Maltiverse					✓				
46	Malware Domain List									✓
47	MalwareDomain									✓
48	Metadefender		✓							
49	MITRE	✓								
31	ExoneraTor		✓							
32	Exploitalert	✓								
33	ZeuS Tracker	✓								
34	FireHOLIP Lists					✓				
35	FraudGuard		✓							
36	Grey Noise		✓							
37	Hail a TAXII									✓
38	HoneyDB									X
39	Icewater	✓								
40	Infosec-CERT-PA	✓								
41	I-Blocklist									✓
42	Majestic Million									✓
43	Malc0de DNS Sinkhole	✓								
44	MalShare.com					✓				
45	Maltiverse					✓				
46	Malware Domain List									✓
47	MalwareDomain									✓
48	Metadefender		✓							
49	MITRE	✓								
69	signature-base									✓
70	SecurityAffairs							✓		
71	Strongarm		✓							
72	SSL Blacklist									✓
73	Symantec		✓							

Table 1.3 Continued.

Sr. No.	Sources	VU DB	Website vendor	List mailing	Blog	Forum	Portal for streaming	News website	Social net	Others
74	Symantec		✓							
75	The Spamhaus Project									✓
76	Talos Aspis		✓							
77	threatfeeds.io									✓
78	Threatglass	✓								
79	ThreatMiner									✓
80	WSTNPHX									✓
81	URLhaus									✓
82	US-CERT							✓		
83	Vulners	✓								
84	VuldB	✓								
85	VirusShare					✓				
86	VirustotaL									✓
87	Yara-Rules									✓
88	YouTube						✓			
89	WeliveSecurity							✓		
90	ZeuS Tracker									✓
91	Phishtank									✓
92	OtxAlienvault					✓				
93	NVD NIST	✓								
94	IntelTechniques									✓
95	Rreddit								✓	
96	Ffsisac							✓		
97	TheHackerNews							✓		
98	CXSecurity	✓								
99	ITS-Security News							✓		
100	GitHub									✓
101	Twitter								✓	
102	TSecurity							✓		
103	BugsChromium					✓				
104	OpenWall									✓
105	LegalHackers				✓					
106	Fortinet				✓					
107	0patch				✓					
108	TrendMicro				✓					
109	KitPloit				✓					
110	NakedSecurity							✓		
111	QuarksLab				✓					
112	Malware-dontneedcoffee							✓		
113	Esecpro				✓					
114	Quttera				✓					
115	IBM		✓							
116	OpenSSL		✓							
117	DailyMotion						✓			
118	MISP									✓
119	CRITS									✓
120	OpenBSD		✓							
121	Microsoft		✓							
122	Mozilla		✓							

(Continued)

Table 1.3 Continued.

Sr. No.	Sources	VU DB	Website vendor	List mailing	Blog	Forum	Portal for streaming	News website	Social net	Others
123	Bugzilla									✓
124	TrendMicro		✓							
125	Googleblog				✓					
126	KrebsonSecurity				✓					
127	ThreatPost							✓		
	Total	**15**	**16**	**4**	**14**	**10**	**2**	**14**	**2**	**50**

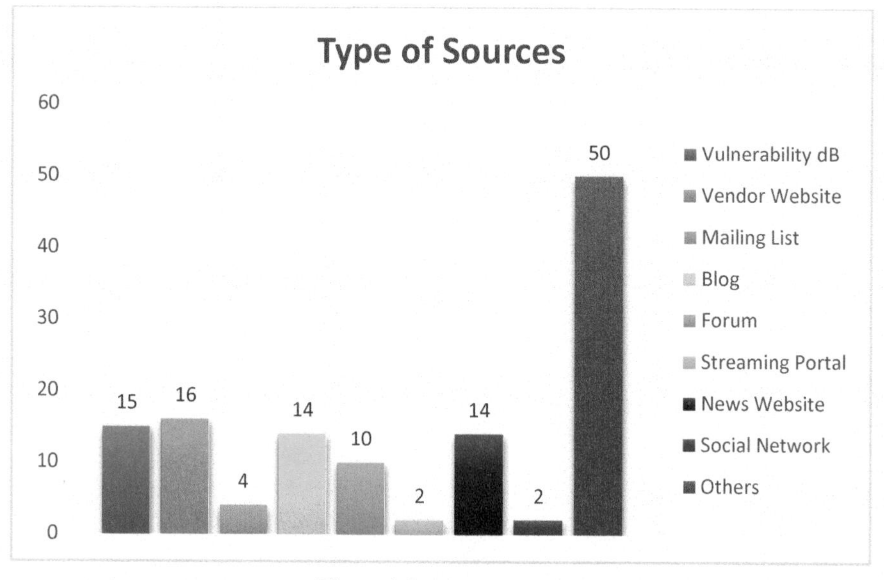

Figure 1.2 Sources type

1.5 Trust and Anonymity in Threat Intelligence Platforms

Trust plays a vital role in sharing cyber threat intelligence. Trusted relationships raise confidence in the stakeholders that the provided information will be used as it is planned for. For example, participants do not create damage with vulnerability information, mainly if they have not been fixed yet. "Furthermore, participants have and, additionally, that stakeholders have suitable safety measures in place and exchange the information as identified. In this regard, it is essential to understand the value that each actor adds to the exchange. Classifying membership conditions for any sharing of data helps create transparency and confidence from day one."

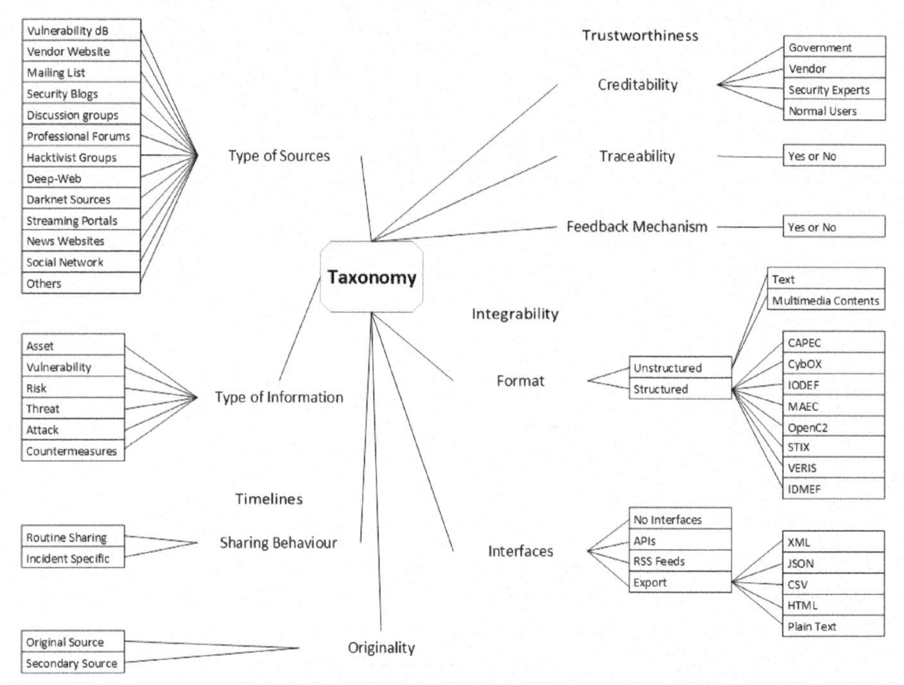

Figure 1.3 Taxonomy for information security data source

i. **Trust:** Platforms build their cyber threat intelligence (CTI), whether cooperation with other stakeholders is permitted, whether platforms are shared only within the industry, and whether outside sources are allowed to drift into the threat intelligence platforms if an internal vetting procedure is intact. "Internal trust established for the stakeholder is labeled with platforms T1. Platforms T2 indicate no trust was established previously and to be initiated manually. Platforms which are indicated with both T1 and T2 and offer a trusted atmosphere for participants but also permit the manual network to other peers and sources outside the trusted atmosphere."

Platforms support inner testing techniques to create a trusted atmosphere but disagree with external links. This creates the drawback of cyber intelligence threat sources. Most threat intelligence providers decide to provide their stakeholders with cyber threat intelligence that is deprived of direct contact with others. About one-third of the providers decided to provide manual trust creation, i.e., stakeholders are encouraged to develop trust manually. About five platforms braced the manual links between stakeholders or the external threat and intelligence nourishes along with their examined cyber intelligence threat. Procedures used to create a trust are most likely vetting and are not clear for the users to see. Therefore, it is recommended that stakeholders trust the threat intelligence platforms with their vetting procedures. The threat intelligence providers must deliver a more energetic trust estimation model, primarily, when they back stakeholder collaboration. There are many free platforms like "AlienVault" and "IBM X-Force Exchange." They allow membership with an e-mail and password. When it comes to sharing information about vulnerabilities, these vetting processes are insufficient. The malicious participants may create profiles to monitor present discussions and threat intelligence that may trigger attacks by an adversary. Moreover, these participants could also feed wrong data into the network and occupy other users to confirm the fake information inserted. This might divert the stakeholders from receiving the real threats.

ii. **Anonymity:** Alien Vault's anonymity function does not identify stakeholders' identity including the system's data and internal IP traffic. If an input is made, external IP addresses, traffic patterns, timestamps, and indicator of compromise (IOC) activity data are collected. So, in this hub-to-spoke model, an anonymity function is provided. Yet, it does not provide any anonymity relating to the content, i.e., removing or masking of personally identifier information (PII) that must be steered manually before sharing CTI. The HP Threat Central platform allows a preprocessing of the data to remove specific information before being shared with the community. The platform encrypts the communication channel over HTTPS. Furthermore, it provides a comprehensive policy setting to decide with whom to share. By contrast, Comilion provides anonymization at the architecture level, making the sender untraceable. This is a valuable and imperative function; nevertheless, the sharing stakeholder could be de-anonymized by analyzing the unmasked personal identified information (PII).

1.6 Time (Speed) in Threat Intelligence Platforms (TAXII)

Some platforms were deployed and tested against the parameter time in the second phase. How quickly they send or receive data from client to server or server to client. Twelve servers and clients were able to be deployed and implemented successfully. One data source was also installed by using its API data and can be used as an input for clients to send to servers or each

Table 1.4 Trust and anonymity in threat intelligence platforms

Platforms	Vetting process	Own cyber threat intelligence	External sources	Trust	Anonymity
Accenture Cyber Intelligence Platform		✓		T1	
Alien Vault Open Threat Exchange (OTX)			✓	T2	
Anomaly Threat Stream (STAXX)			✓	T2	
Anubis Networks Cy-buffered		✓		T1	
Center (R-CISC) intelligence sharing portal		✓		T2	
Cisco Talos		✓		T1	
Comilion			✓	T2	
Crowd Strike Falcon Platform		✓		T1	
Cyber Connector	✓	✓		T1	
Cybersecurity Information Sharing Partnership	✓			T1	
Defense Security Information Exchange	✓			T1	
Eclectic IQ	✓	✓	✓	T1, T2	
Facebook Threat Exchange	✓			T1	
Health Information Trust Alliance Cyber Threat Xchange (CTX)	✓			T1	
HP Threat Central			✓	T1	
IBM X-Force Exchange			✓	T2	
Infoblox Threat Intelligence Data Exchange		✓		T1	
Last Quarter Mile Toolset (LQMT)		✓		T1	
Looking Glass Scout Prime (Cyveillance)		✓	✓	T1, T2	
Malware Information Sharing Platform (MISP)			✓	T2	
McAfee Threat Intelligence Exchange		✓		T1	
Microsoft Interflow			✓	T2	
NC4 CTX/SoltraEdge			✓	T2	
NECOMAtter (NECOMAtome)			✓	T2	
Norm Shield		✓		T1	
Recorded Future		✓		T1	
Retail Cyber Intelligence SharingCenter (R-CISC) Intelligence Sharing Portal			✓	T2	

(Continued)

Table 1.4 Continued.

Platforms	Vetting process	Own cyber threat intelligence	External sources	Trust	Anonymity
ServiceNow-Bright Point Security		✓		T1	
Splunk		✓	✓	T1	
Threat Connect			✓	T1	
Threat Track Threat IQ		✓	✓	T1, T2	
Threat Quotient	✓	✓	✓	T1, T2	

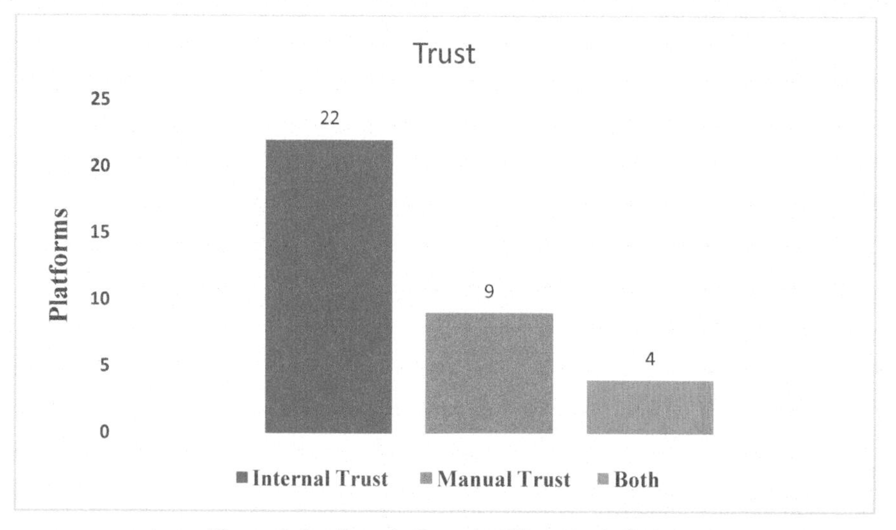

Figure 1.4 Trust in threat intelligence platforms

other. All platforms accept data in different formats, making testing more challenging. So there is no choice to test only those that accept data from our data generator, which supports format STIX 2.0.

i. **OpenTAXII** is a Python implementation of TAXII Services 1.0 and 1.1 which delivers through its API message exchanges to share cyber threat intelligence information between parties. It is also no longer in use and is unable to be used in the production environment.
ii. **Cabby** is a command-line tool and an open-source Python library. It provides the developers with easy support for interconnecting TAXII Services 1.1 and 1.0. It is also no longer in use and is unable to be used in a production environment.

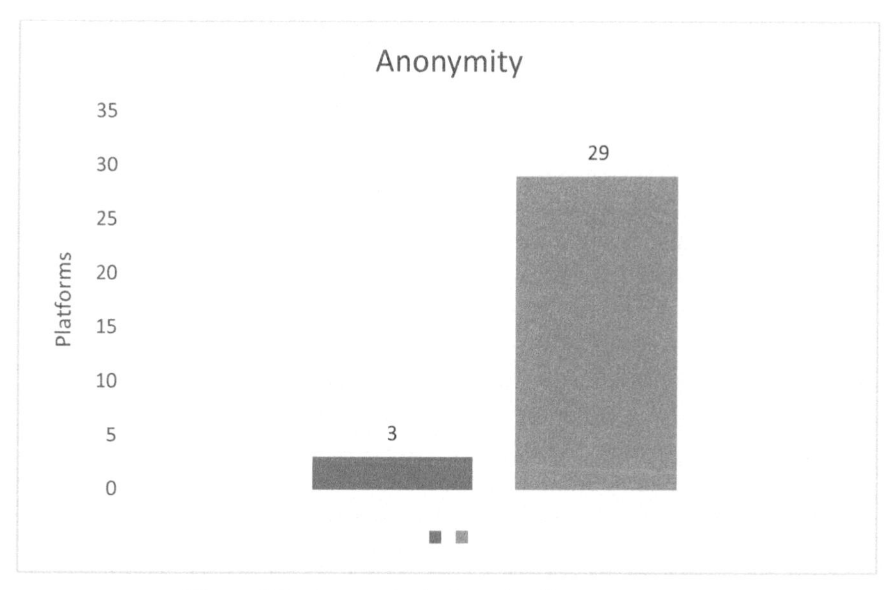

Figure 1.5 Anonymity in threat intelligence platforms

iii. **CTI TAXII Server 2.0** is TAXII 2.0 implementation designed in Python framework Flask. Medallion has been designed to give a simple front-end REST server providing access to the endpoints. TAXII server stores STIX 2 content. It gives the Memory backend and the MongoDB back-end. The Memory backend persists data "in memory." It initialized using a JSON file that contains TAXII data and metadata. The MongoDB backend is somewhat more robust and uses a MongoDB server, installed independently. It is easy to install and use with some configurations external STIX content can be used as an input and sent on the server. From all servers, it is the best and can be used in production. It contains fewer fields and data in JSON STIX format, making it faster. Within fewer time, data can be shared and accessed. Also, it is developed by OASIS.

iv. **OASIS** is a nonprofit consortium that drives the development, convergence, and adoption of open standards for the global information society. OASIS promotes industry consensus and produces worldwide standards for security, Internet of Things, cloud computing, energy, content technologies, emergency management, and other areas. OASIS open standards offer the potential to lower cost, stimulate innovation, grow global markets, and protect the right to free technology choice.

v. **CTI TAXII Client 2.0** is a client implementation for the TAXII 2.0 Server. It supports the following TAXII 2.0 API services: server discovery, get API root information, get status, get collections, get a collection, get objects, add objects, get an object, and get object manifests. It integrates with the server and makes it faster to access or write data. Free TAXII 2.1 is a TAXII 2.1 Server written in the Go language. It supports STIX 2.1. It is also a suitable server for sharing threat intelligence, but it only supports data from files. It does not save data in the database and only gives static data. This limitation makes it less usable. And, second, it took time and some effort to configure this server.

vi. **SRA TAXII Server 2.0:** This is also a good server, but it is challenging to set up plus there is no client with it, which makes it more difficult to access data. It provides API for data extraction but not for sending data to the server. Due to its limitations, it is least useful.

vii. **Anomaly STAXX** is a good full-fledge TAXII client, but it has its data format and has limitations only gathering information from its own defined data sources that required license to use external data sources. In the free version, it only provides test data which comes from its servers, i.e., LIMO Server.

viii. **MISP** is a good open-source threat intelligence sharing platform that is expandable. It can import many types of formats and export them into many formats. We send and receive data from MISP, but it is slower than CTI TAXII Server 2.0. The reason is that it contains many attributes. It can easily integrate with other threat intelligence sharing platforms, such as NIDS and LIDS, and also log analysis tools, such as SIEMs. We also integrate it with our TAXII server to send MISP data through the TAXII server to another end.

ix. **SoltraEdge**, **TripWire**, and **Splunk** come with trial and support data of their formats, making it incompatible with standard data format, in our case, STIX 2.0. This limitation is only in free trial versions. They are set up and installed on the buyer's behalf in paid premium versions.

1.7 Receiving Time in Threat Intelligence Platforms (TAXII)

All values are average and may change depending on the traffic on the network. In the third phase, sources of threat, we present the scheme of classification that has been created and the results of the assessment data sources of information security.

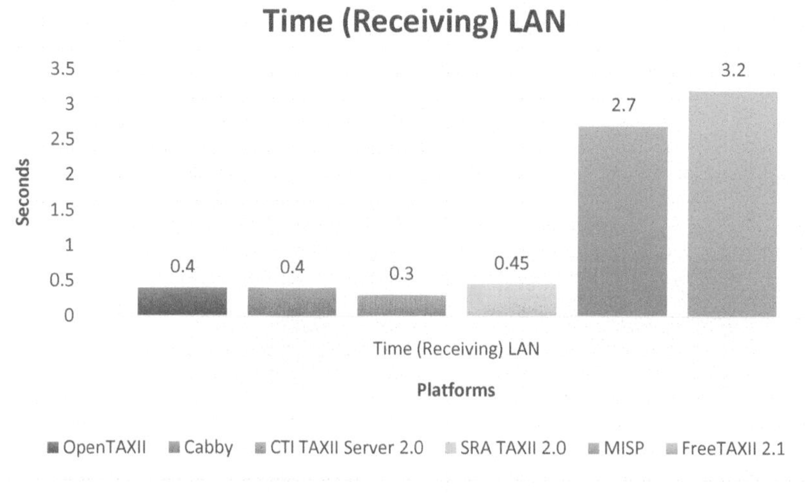

Figure 1.6 Time receiving (LAN) in threat intelligence platforms (TAXII)

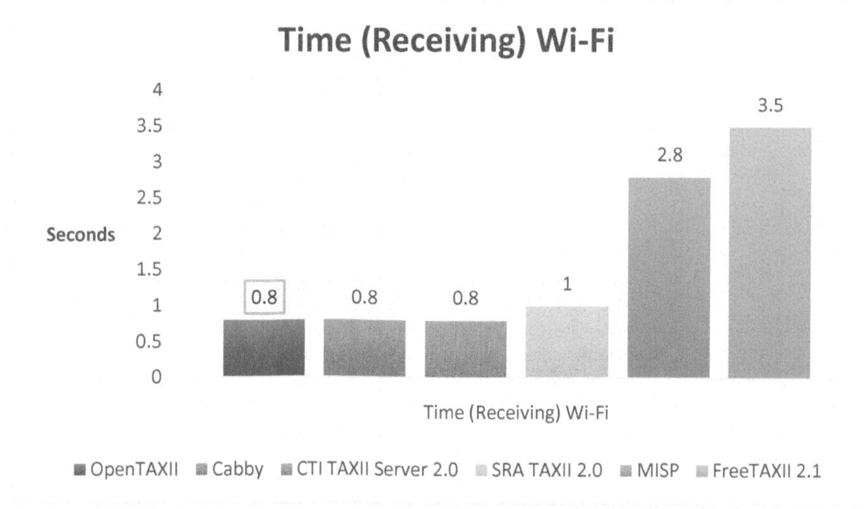

Figure 1.7 Time receiving (Wi-Fi) in threat intelligence platforms (TAXII)

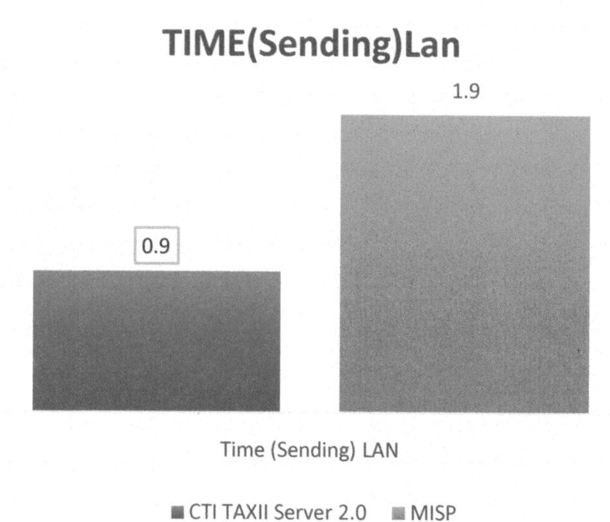

Figure 1.8 Time sending (LAN) in threat intelligence platforms (TAXII)

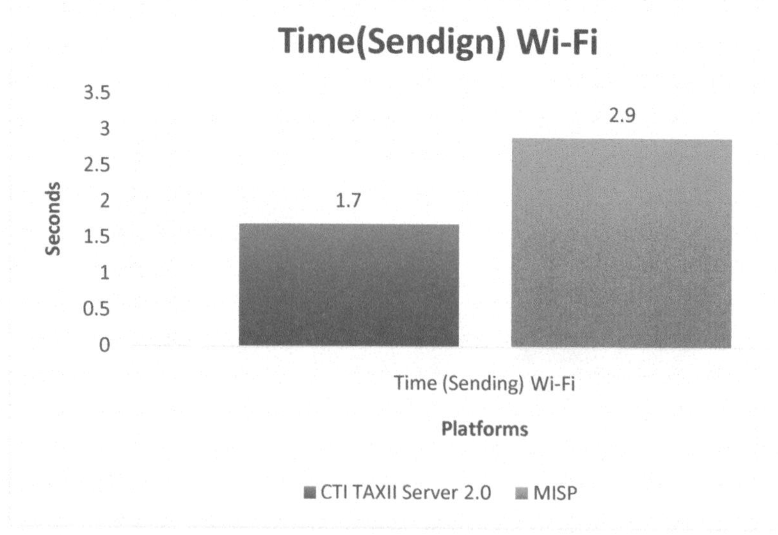

Figure 1.9 Time sending (Wi-Fi) in threat intelligence platforms (TAXII)

1.8 Conclusion

In a nutshell, research provides a comprehensive comparative analysis of threat intelligence sharing platforms and structure classification for the data sources of information security used in research practice and analyzed approximately 127 data sources for analysis in our research methodology. To classify these data sources, a taxonomy was created based on information type, timeliness, integrability, originality, source type, and trustworthiness features. For the classification of 127 data sources of information security, a proper taxonomy is used. It was found that most of the sources concentrate on information related to vulnerabilities. Our result also showed that the data is unorganized, missing or inadequate interfaces restrict automated integration, and some sources acquire and duplicate the data from popular data sources of information security. Another platform is compared based on trust, anonymity, and time parameters. Trust plays a vital role in sharing cyber threat intelligence. Trusted relationships raise confidence in the stakeholders that provide information that will be used as planned for trust linkages amongst sharing stakeholders are imperative to exchange cyber intelligence threats. However, finding the correct elements to create a trusted atmosphere is difficult. This study evaluated 32 threat intelligence providers for trust functionality and anonymity. Based on our research, MISP is the best and an open-source, expandable platform for threat intelligence sharing which can be further extended to enhance the capability, trust, and maturity of threat intelligence sharing model.

References

[1] J. Jang-Jaccard and S. Nepal. A survey of emerging threats cybersecurity. *Journal of Computer and System Sciences*, 80:973–993, 2014.

[2] P. Puhakainen and M. Siponen. Improving employee's compliance through information systems security training: action research. *Management Information Systems*, 757–778, 2010.

[3] Z. A. Soomro, M. H. Shah, and J. Ahmed. Information security management need a more holistic approach: A literature review. *International Journal of Information Management*, 36:215–225, 2016.

[4] Y. Harel, I. B. Gal, and Y. Elovici. Cybersecurity and the role of intelligent systems in addressing its challenges. *ACM Trans Intell Syst*, 2017.

[5] C. Sauerwein, M. Gander, M. Felderer, and R. Breu. A systematic literature review of crowdsourcing-based research in information security. *Proceedings of the 2016 IEEE Symposium on Service-Oriented System Engineering (SOSE), IEEE*, pages 364–371, 2016.

[6] C. Sauerwein, C. Sillaber, A. Mussmann, and R. Breu. Threat intelligence sharing platforms: An exploratory study of software vendors and research perspectives. 2017.

[7] V. Mavroeidis and S. Bromander. Cyber threat intelligence model: An evaluation of taxonomies, sharing standards, and ontologies within cyber threat intelligence. *Intelligence and Security Informatics Conference (EISIC)*, pages 91–98, 2017.

[8] J. Steinberger, A. Sperotto, M. Golling, and H. Baier. How to exchange securityevents? Overview and evaluation of formats and protocols. *IEEE International symposium on Integrated Network Management (IM), IEEE*, pages 261–269, 2015.

[9] J. L. Hernandez-Ardieta. J. E. Tapiador, and G. Suarez-Tangil. Information sharing models for cooperative cyber defense. *Proceedings of the 2013 5th International Conference on Cyber Conflict*, pages 1–28, 2013.

[10] S. Qamar, Z. Anwar, M. A. Rahman, E. Al-Shaer, and B. Chu. Data-driven analytics for cyber-threat intelligence and information sharing. *Computers and Security*, 2017.

[11] T. Mattern, J. Felker, R. Borum, and G. Bamford. Operational levels of Cyber Intelligence, 27, 2014.

[12] E. W. Burger, M. D. Goodman, P. Kampanakis, and A. Z. Kevin. Taxonomy model for cyber threat intelligence information exchange, *ACM Workshop on Information Sharing and Collaborative Security*, 2014.

[13] W. Tounsi and H. Rais. A survey on technical threat intelligence in the age of sophisticated cyber attacks. *Computers and Security*, 2017.

[14] F. Menges and G. Pernul. A comparative analysis of incident reporting formats. *Computers and Security*, 2017.

[15] H. Dalziel. *How to Define and Build an Effective Cyber Threat Intelligence Capability*. Waltham, MA, USA: Syngress, 2015.

[16] I. Gartner. Threat intelligence: What is it, and how can it protect you from today's advanced cyber-attacks?, 2014.

[17] J. Connolly, M. S. Davidson, M. Richard, and C. W. Skorupka. *The Trusted Automated Exchange of Indicator Information*. McLean, VA, USA: The Mitre Corporation, 2014.

[18] S. Barnum. Standardizing cyber threat intelligence information with the structured threat information expression (stixtm). McLean, VA, USA: MITRE Corporation, 2012, p. 11.

[19] L. Dandurand and O. Serrano. Towards improved cybersecurity information sharing. *5th International Conference on Cyber Conflict (CyCon), IEEE*, pages 1–16, 2013.

[20] S. Brown, J. Gommers, and O. Serrano. From cyber security information sharing to threat management. *2nd ACM Workshop on Information Sharing and Collaborative Security, ACM*, pages 43–49, 2015.

[21] O. Serrano, L. Dandurand, and S. Brown. On the design of a cybersecurity data sharing system. *ACM Workshop on Information Sharing and Collaborative Security, ACM*, pages 61–69, 2014.

[22] S. Appala, N. Cam-Winget, D. McGrew, and J. Verma. An actionable threat intelligence system using a publish-subscribe communications model. *2nd ACM Workshop on Information Sharing and Collaborative Security, ACM*, 2015.

[23] P. Kampanakis. Security automation and threat information-sharing options. *IEEE Security and Privacy*, 12:42–51, 2020.

[24] R. Martin. Making security measurable and manageable. *IEEE Military Communications Conference, IEEE*, pages 1–9, 2009.

[25] P. Poputa-Clean. Automated defense - using threat intelligence to augment security. Rockville, MD, USA: SANS Institute InfoSec, 2015.

[26] T. D. Wagner (B), E. Palomar, K. Mahbub, and A. E. Abdallah. *Towards an Anonymity Supported Platform for Shared Cyber Threat Intelligence*. Birmingham, U.K.: Birmingham City University, 2018.

[27] T. D. Wagner (B), E. Palomar, K. Mahbub, and A. E. Abdallah. A novel trust taxonomy for shared cyber threat intelligence. E. Vasilomanolakis (Ed.), 2020.

[28] M. S. Abu, S. R. Selamat, A. Ariffin, and Y. Robiah. Cyber threat intelligence - issue and challenges. *Indonesian Journal of Electrical Engineering and Computer Science*, 10:371–379, 2018.

[29] C. Johnson, L. Badger, D. Waltermire, J. Snyder, and C. Skorupka. *Guide to Cyber Threat Information Sharing*. Gaithersburg, MD, USA: NIST, 2016.

[30] C. Sillaber, A. Mussmann, C. Sauerwein, and R. Breu. Data quality challenges and future research directions in threat intelligence sharing practice, WISCS '16: Proceedings of the 2016 ACM on Workshop on Information Sharing and Collaborative Security, 2016.

[31] C. Sauerwein, C. Sillaber, and R. Breu. Shadow cyber threat intelligence and its use in information security and risk management processes, *Multikonferenz Wirtschaftsinformatik*, 2018.
[32] C. Sauerwein, I. Pekaric, M. Felderer, and R. Breu. An analysis and classification of public information security data sources used in research and practice. *Computers & Security*, 82:140–155, 2019. doi: 10.1016/j.cose.2018.12.011.URL https://doi.org/10.1016/j.cose.2018.12.011.
[33] C. Cappiello, C. Francalanci, A. Maurino, and C. Batini. Methodologies for data quality assessment and improvement. *ACM Computing Surveys*, 2009.
[34] S. J. Kuhnert, B. Sperotto, A. Baier, and H. A. Pras. Cyber threat intelligence issue and challenges in whom do we trust-sharing security events. *Proceedings of the IFIP International Conference on Autonomous Infrastructure, Management and Security, Springer*, 10:24–111, 2016.
[35] National Vulnerability Database (NVD). 2019. URL https://nvd.nist.gov/.
[36] J. McHugh, W. A. Arbaugh, and W. L. Fithen. Windows of vulnerability: A case study analysis. *Computer*, 13:9–52, 2000.
[37] B. Plattner, S. Frei, and B. Tellenbach. *0-Day Patch-Exposing Vendors (In)Security Performance*. London, U.K.: BlackHat Europe, 2008.
[38] G. White and W. Zhao. A collaborative information sharing framework for community cyber security. *Proceedings of the 2012 IEEE Conference on Technologies for Homeland Security (HST), IEEE*, pages 62.
[39] J. Baker. Taxii client, 2019. URL https://github.com/TAXIIProject/yeti.
[40] EclecticIQ. Taxii server implementation in python from eclecticiq, 2019.
[41] EclecticIQ. Taxii client implementation from eclectic, 2019. URL https://github.com/eclecticiq/cabby.
[42] OASIS TC. Taxii 2 server library written in python, 2019. URL https://github.com/oasis-open/cti-taxii-server.
[43] OASIS TC. Taxii 2 client library written in python, 2019. URL https://github.com/oasis-open/cti-taxii-server.
[44] OASIS TC. Lightweight visualization for STIX 2.0 objects and relationships, 2019. URL https://github.com/oasis-open/cti-stix-visualization.
[45] B. Jordan. A cyber threat intelligence server based on taxii 2 and written in Golang, 2019. URL https://github.com/freetaxii/server.
[46] Security Risk Advisors. Taxii 2.0 server implemented in node JS with Mongodb backend, 2019. URL https://github.com/SecurityRiskAdvisors/sra-taxii2-server.

[47] MISP. Misp - open source threat intelligence platform and open standards for threat information sharing, 2019. URL https://www.circl.lu/misp-images/latest/.

[48] Soltra. SoltraEdge, 2019. URL https://updates.soltra.com/download.php.

[49] Splunk Enterprise 7.2.6. Splunk enterprise 7.2.6, 2019. URL https://www.splunk.com/en_us/download/splunk-enterprise.html.

2

Evaluation of Open-source Web Application Firewalls for Cyber Threat Intelligence

Oumaima Chakir[1], Yassine Sadqi[1], and Yassine Maleh[2]

[1]Laboratory LIMATI, FPBM, USMS University, Morocco
[2]Laboratory LaSTI, ENSAK, USMS University, Morocco
E-mail: chakir.oumaima@hotmail.com; .sadqi@gmail.com; .maleh@ieee.org

Abstract

One of the most well-known attack detection and prevention systems for web application security is the web application firewall. In this chapter, we present a study on the effectiveness of the most popular and widely used open-source web application firewalls, named AQTRONIX Webknight v4.4 and ModSecurity v3.0.4, within the four paranoia levels of CRS v3.3.0. According to the experimental results based on the Payload All The Thing and CSIC HTTP 2010 datasets, AQTRONIX Webknight is an effective system for securing web applications against attacks, having identified all the attacks launched with a recall value of 98.5%. It has, however, produced a high false positive rate, with a value of 99.6%. Furthermore, ModSecurity capability is dependent on the PL of CRS configured. We discovered that as PL increases, so does the number of recognized assaults. Unlike the other levels, ModSecurity had a high false positive rate at level 4, with a value of 60.3%.

Keywords: Web application security, attack, web application firewall (WAF), ModSecurity, core rule set (CRS), AQTRONIX Webknight, Payload All The Thing, CSIC HTTP 2010 dataset.

2.1 Introduction

Day by day, the internet plays a significant role in our lives, and numerous web applications have been implemented in various fields such as education, healthcare, business, and social networks. Generally, a web application can be defined as a software hosted on a remote server and accessible via a network using a browser client [1], [2].

As a consequence of their popularity, they have become an attractive target for the attacker's community, who aggressively exploit web application vulnerabilities [3], [4]. In general, attacks against web applications aim to mimic normal user behavior to damage the web apps and/or steal sensitive information, such as the user password, financial, and payment information [5], [6]. In this context, the use of a web-based attacks detection system becomes necessary for web application security.

Traditional security systems, such as firewalls and intrusion detection and prevention systems (IDPS), are primarily dedicated to network security and ineffective for protecting web applications since the nature of attacks on the web is very different from that on the network [7].

Traditional firewalls (stateless and stateful) are systems that operate at the lower layers of the OSI model, up to layer 4. So, when exploiting a vulnerability of a web application (OWASP Top 10 [8]), HTTP traffic will be allowed by this system since it is unable to analyze its content. Intrusion detection and prevention systems are still immature in the field of web application security. According to [9], the design of a suitable IDPS to prevent web attacks still needs to be examined more closely.

Web application firewall, for short WAF, is one of the most popular and used systems that can ensure the security of a web application in the back end [7], [10]. Web application firewall hardware or software is a monitoring system that can analyze, detect, and prevent malicious HTTP/s traffic in real time, without affecting the performance of the web application [11], [12]. Before it reaches the web application, WAF evaluates each HTTP/s request to ensure that it conforms to the defined security policy. Same for HTTP/s responses, it handles responses sent by the application to prevent leakage of sensitive data via error messages. It can be deployed in reverse proxy mode, acting as an intermediary between the client and the web server, which allows hiding the infrastructure hosting the web application, in integrated mode or in SPAN mode, where the WAF is unable to prevent malicious HTTP/s requests.

In general, the SPAN mode is designed to guarantee that the WAF is working effectively. To detect attacks, a WAF supports different detection approaches: signature, anomaly, and hybrid detection approach [9], [11], [13].

Signature-based WAFs cannot detect attacks that are not explicitly expressed by signatures. On the other hand, anomaly-based WAFs can detect both known and unknown attacks; they consider that any deviation from the reference behavior is harmful [11], [13], [14]. Hybrid-based WAFs combine several detection approaches. These combinations make it possible to benefit from the advantages of each approach used and to overcome their disadvantages [11], [13].

WAFs have many features. Among them, we can mention deep analysis of the contents of the HTTP/s header and request body to verify the presence or absence of an attack, protection against the most common vulnerabilities listed in the OWASP Top 10 such as XSS, injection attack, etc. Virtual vulnerability correction is also an important feature. WAF can be used as a virtual patching solution to prevent the exploitation of newly discovered vulnerabilities.

Motivated by the popularity of these systems, we investigate the effectiveness of the most popular open-source WAFs, named ModSecurity [15] and AQTRONIX Webknight [16] in web application security. These WAFs were chosen because of their high community support and rich documentation.

The contributions of this work are summarized as follows:

- Theoretical study of the most widely used open-source WAFs.
- Experimental evaluation of open source WAFs: ModSecurity v3.0.4 with CRS v3.3.0 and AQTRONIX webknight v4.4 using CSIC HTTP 2010 dataset.
- Experimental study on the effectiveness of ModSecurity, AQTRONIX Webknight in detecting SQL Injection (SQLI), Reflected Cross-Site Scripting (XSS), XML External Entities (XXE) attacks using Payloads All The Things.

This chapter is structured as follows. Section 2.2 presents a theoretical study of ModSecurity and AQTRONIX Webknight. The research methodology is discussed in detail in section 2.3. The experimental results are presented and discussed in Section 2.4. Section 2.5 presents some recommendations for developing secure web applications, and Section 2.6 is dedicated to the conclusion and future works.

2.2 Open-source Web Application Firewalls

2.2.1 ModSecurity

In 2002, Ivan Ristic created one of the most powerful and commonly used web application firewalls called ModSecurity. It is an open-source web application firewall that analyzes HTTP/s traffic entering and departing the web application to detect and block known attacks based on a set of rules known as the OWASP Core Rule Set [13], [15], [17].

The CRS is an OWASP-created list of known attack signatures designed to improve ModSecurity detection capability. It has four paranoia levels: PL1 (the default level), PL2, PL3, and PL4. The number of rules increases in proportion to the PL. It aims to protect web applications with a minimum of false alerts against a wide range of attacks, including the OWASP Top 10 [18].

According to [17], ModSecurity has five processing phases, each one corresponding to a critical step. The first four phases are for processing the header and body of the request and response, while the last phase is for logging triggered events. For each transaction, the phases are executed sequentially from phase 1 to phase 5.

Ivan Ristic describes the role of each phase in [17] as follows:

- **Phase 1: Request Headers**
  ModSecurity evaluates the content of the request header in this phase, and only the rules defined in that phase are loaded. ModSecurity proceeds to phase 2 if no rule fits the examined input. Otherwise, if a rule triggers a match, it notifies the webserver and records the event (phase 5).
- **Phase 2: Request Body**
  Similarly, ModSecurity examines the content of the request body and, when it detects the presence of an attack, i.e., a rule in this phase that triggers a match with the reviewed content, it blocks and logs it. Otherwise, no correspondence, ModSecurity transmits the request to the webserver.
- **Phase 3/4: Response Headers/Body**
  The headers and bodies of HTTP responses are also observed by ModSecurity to prevent the leakage of information via error messages. The processing of these two phases is identical to phases 1 and 2.
- **Phase 5: Logging**
  The rules defined at this level only impact how logging should be performed. ModSecurity is capable of saving all the data of a transaction, and we can also specify which part of the transaction should be saved.

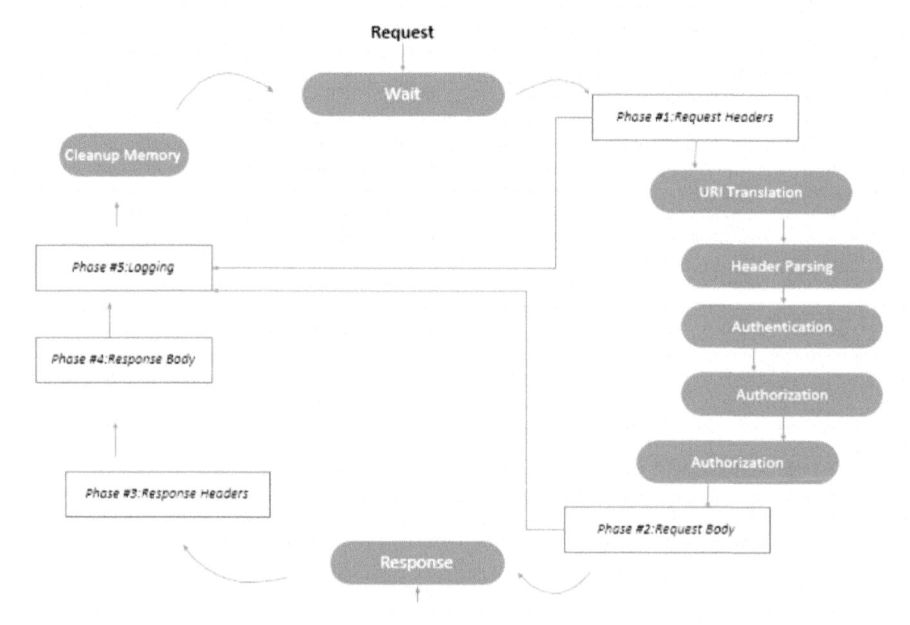

Figure 2.1 The five phases of ModSecurity analysis [13]

2.2.2 AQTRONIX Webknight

AQTRONIX Webknight is an open-source web application firewall for IIS web servers, developed by AQTRONIX and released under the GNU license. It analyzes all requests and processes them according to the filtering rules defined by the administrator.

According to [16], Webknight examines the HTTP request headers, verbs, URL, HTTP version, queries, and entity data sent to the web application to detect attacks, such as SQLi, XSS, CSRF/XSRF, DoS attacks, and so on.

In [9], Sadqi and Mekkaoui evaluated the performance of four of the most open-source intrusion detection and prevention system (IDPS), such as AppSensor, ModSecurity, Shadow Daemon, and AQTRONIX Webknight based on various security features that must be incorporated into a web security system.

In Table 2.1,we present an assessment of ModSecurity and AQTRONIX Webknight based on Sadqi and Mekkaoui's [9] work.

As shown in Table 2.1, ModSecurity and Webknight are unable to detect unknown attacks as they are both based on the signature detection approach.

Concerning the input validation and response-time criteria, the aforementioned WAFs can verify the validity of the data entered by the user

Table 2.1 Assessment of ModSecurity and AQTRONIX Webknight

Functionality	ModSecurity	Webknight
Detection method	Signature-based	Signature-based
Input validation	Yes	Yes
Output sanitization	No	No
Session verification	No	No
Access control	No	Yes
Bots detection	No	Yes
Response-time	Real-time	Real-time
Placement	Reverse-proxy, Server-side	Server-side

and reply in real time without affecting the application's performance. Only ModSecurity can ensure the validation of outputs and session verification.

For robots detection, AQTRONIX Webknight is the best solution because it has a large database of robots which allows it to block or authorize only certain types of robots, set up a trap against malicious robots, and block aggressive robots.

2.3 Research Methodology

Evaluation of intrusion detection and prevention systems is critical in the field of web security. It provides recommendations and information to assist administrators in selecting the best and most appropriate protection mechanism for the security of their web applications. In this chapter, we investigate the effectiveness of the most popular open-source web application firewalls, ModSecurity v3.0.4 and AQTRONIX Webknight v4.4. We have divided the experiment into two phases.

In the first phase, we tested the ability of ModSecurity v3.0.4 using CRS v3.3.0 and AQTRONIX Webknight v4.4 in detecting three types of attacks, SQLI, XSS, and XXE, using Payloads All The Things [19] as a dataset. In the second phase, we evaluated the performance of these WAFs using CSIC HTTP 2010 dataset [20].

2.3.1 Implementation of ModSecurity and AQTRONIX Webknight

ModSecurity and Webknight are WAFs that can be deployed in reverse proxy or integrated mode. In our case, we chose to implement the WAFs in integrated mode, which means that we installed and configured the target

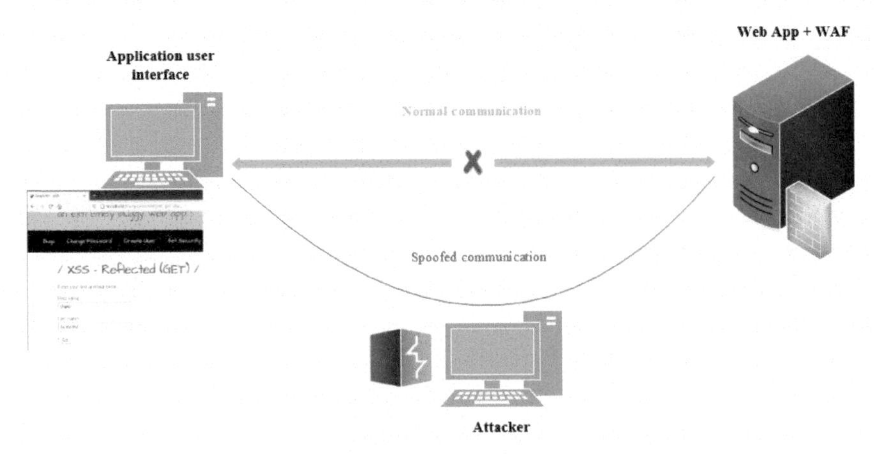

Figure 2.2 Attack scenario

application, the attack tool, and the WAF on the same machine. Furthermore, the default configuration of these WAFs has not been changed.

As can be seen from Figure 2.2, we need three components. The first is bWAPP [21] which is the target application that we want to protect, the second is the WAF, and the last is Burp Suite [22] that will play the role of an attack launcher. From the target, we send HTTP requests and we intercept them with Burp Suite to modify them and inject malicious and legitimate data, and then we resend them to see how our WAF will react.

2.3.2 Dataset Description

2.3.2.1 Payload All The Thing

is a list of payloads used to bypass the security of web applications released on GitHub in 2018. It contains payloads for the most known vulnerabilities exploited by attackers, such as SQLI, XSS, XXE, Blind injection, and so on [19].

CSIC HTTP 2010 is a dataset that contains real requests generated to an e-commerce application. It includes various types of web attacks such as SQL Injection, Cross-Site Scripting, CRLF Injection, and Buffer overflow [20].

Table 2.2 Number of queries in each payload for each attack

Payload	SQLI	Reflected XSS	XXE
No. of Requests	672	667	67

Table 2.3 CSIC HTTP 2010 Dataset Size

Dataset	Train valid	Test valid	Test attack
CSIC 2010	36,000	36,000	25,065

Table 2.4 Tools used in this work

Virtual machine	Virtual box 6.1.12
Victim application	bWAPP
Attacker	Burp Suite v2020.9.2
OWASP ModSecurity CRS version	3.3.0
OS	Ubuntu 18.04
	Windows 7 64bits
Servers	Apache 2.4.29
	IIS server
WAFs	ModSecurity v3.0.4
	AQTRONIX Webknight v4.4

To evaluate the aforementioned WAFs, we used the test data of this dataset. Among 25,065 attacks and 36,000 legitimate requests, 3973 attacks and 1517 legitimate requests were used.

2.3.3 Experiment Environment

To complete this work, we utilized a PC-equipped Intel(R) Core (TM) i5-3320M*64 based CPU running at 2.6 GHZ with 12 GB RAM and Windows 10 64 bit as an operating system. The various tools used to carry out this work are summarized in Table 2.4.

bWAPP: An open-source web application that is voluntarily insecure developed in PHP and uses a database MySQL. It is possible to host it under Windows or Linux with Apache or IIS. It is intended to help web developers, educators, and students learn and improve their skills.

bWAPP has more than 100 web vulnerabilities that cover the most common known bugs such as OWASP Top 10, HTML and LDAP injection, command and blind injection, and so on. It has three levels of security: the low level is not protected, the medium level is protected but still vulnerable, and the high level is secure [21].

Burp Suite: To simulate the presence of an attacker, we chose to use the Burp Suite, a powerful software developed by PortSwigger which allows us to intercept and manipulate messages exchanged between two machines. There are three versions of the Burp Suite: the professional version, the enterprise version, and the community version which is a free version [22].

Two tools were used to complete this work:

- Burp Proxy: To intercept and modify all requests we sent from the browser and receive the response from the server.
- Burp intruder: Powerful intrusion tool to automate attacks on web applications.

2.3.4 Evaluation Metrics

The confusion matrix was used to compare the activities performed with the reactions of the mentioned WAFs. From this matrix, we calculated the following metrics:

- Recall (R): Calculates the number of attacks accurately recognized by the WAF.

$$R = \frac{\mathrm{TP}}{\mathrm{TP} + \mathrm{FN}}. \tag{2.1}$$

- Precision (P): Calculates the number of valid requests accurately identified by the WAF.

$$P = \frac{\mathrm{TP}}{\mathrm{TP} + \mathrm{FP}}. \tag{2.2}$$

- F-value ($F1$): The harmonic average of recall and precision.

$$F1 = 2.\frac{\mathrm{R.P}}{\mathrm{R} + \mathrm{P}}. \tag{2.3}$$

- False positive rate (FPR): Number of legitimate requests classified as an attack by the WAF.

$$\mathrm{FPR} = \frac{\mathrm{FP}}{\mathrm{FP} + \mathrm{TN}} \tag{2.4}$$

2.4 Results and Discussion

2.4.1 Results

Tables 2.5–2.8 illustrate the ability of AQTRONIX Webknight and ModSecurity to identify SQLI, XSS, and XXE assaults based on Payload All The Thing.

Table 2.5 SQLI, XSS, and XXE attacks detection using AQTRONIX Webknight

Attack	Detected	Bypassed	Recall
XSS Reflected	667	0	100%
SQLI	672	0	100%
XXE	67	0	100%

Table 2.6 SQLI attacks detection using ModSecurity WAF

SQLI Attack	Detected	Bypassed	Recall
PL1	360	312	53.5%
PL2	461	211	68.6%
PL3	462	210	68.7 %
PL4	464	208	69%

Table 2.7 XXE attacks detection using ModSecurity WAF

XXE Attack	Detected	Bypassed	Recall
PL1	56	11	83.5%
PL2	58	9	86.5%
PL3	61	6	91%
PL4	61	6	91%

Table 2.8 Reflected XSS attack detection using ModSecurity WAF

Reflected XSS	Detected	Bypassed	Recall
PL1	585	82	87.7%
PL2	594	73	89%
PL3	596	71	89.3%
PL4	598	69	89.3%

As shown in Table 2.5, AQTRONIX Webknight is an efficient solution for defending web applications from SQLI, XSS, and XXE attacks, as it completely prevented all attempts with a 100% recall rate.

Unlike Webknight, ModSecurity failed to prevent HTTP requests containing SQLI, XSS, and XXE attacks in all four levels of paranoia.

The efficiency of ModSecurity is highly reliant on the chosen paranoia level, as seen Tables 2.6–2.8 below. When we increased the PL, we observed that the number of detected attacks increased.

Table 2.6 presents the number of SQLI attacks detected in each level of paranoia. In PL1, we observed that 360 of the launched attacks were correctly classified as malicious with a recall value of 53.5%, whereas 312 were classified as valid requests. However, we observed a surge in the number of identified assaults in PL4 with a recall value of 69%.

Table 2.9 Performance evaluation of ModSecurity based on CSIC HTTP 2010 dataset

PL	R	P	F	FPR
PL1	22.2%	100%	36.3%	0
PL2	35.1 %	97.2%	51.5%	2.6%
PL3	74.1%	83.9%	78.6 %	37%
PL4	92.4%	80%	85.7%	60.3%

Table 2.10 Performance evaluation of AQTRONIX Webknight based on CSIC HTTP 2010 dataset

WAF	R	P	F	FPR
Webknight	98.5%	72.1%	83.2%	99.6%

The experimental results based on the CSIC HTTP 2010 dataset proved that the efficiency of ModSecurity is highly dependent on the level of paranoia configured.

Among 3973 attacks, ModSecurity identified 885 attacks in PL1 with a recall value of 22.2%, 1395 attacks in PL2, 2945 attacks in PL3, and 3673 assaults in PL4 with a recall value of 92.4%.

Compared to the other levels, ModSecurity generated a higher FPR in PL4 with a value of 60.3%. We launched 1517 legitimate requests, and 915 requests were classified as malicious.

Unlike ModSecurity, AQTRONIX Webknight reached good performance in terms of accuracy with a value of 98.5%. However, in terms of FPR, AQTRONIX webknight is not accurate since it classified 1512 legitimate requests as malicious with an FPR value of 99.6%.

2.4.2 Discussion

Based on the theoretical and experimental evaluation of ModSecurity and AQTRONIX Webknight, each of these WAFs has advantages and disadvantages. Both of these WAFs are competent in analyzing, detecting, and preventing malicious data in real time, including SQLI, XSS, and XXE attacks. However, they cannot identify assaults that do not have signatures, such as zero-day attacks.

In addition to input validation, ModSecurity can ensure output validation and prevent the leakage of sensitive data, ensuring the verification of user sessions. In contrast, AQTRONIX Webknight is a WAF that can distinguish between requests issued by humans and those delivered by software.

Furthermore, the detection efficiency of ModSecurity is highly dependent on the configured paranoia level (PL). It is advisable not to use PL4, because it has a higher false positive rate than other levels. On the other hand, Webknight is a good solution for detecting known attacks, but it has a higher FPR.

2.5 Recommendations

In this section, we present some recommendations for developing secure web applications:

- The security of web applications must be considered at all stages of development: at design, during development, and so on.
- After the construction of a web application, it is necessary to perform intrusion tests to verify and identify the risks that the application might encounter.
- Error control and management is a critical step to ensure that sensitive information is not displayed to the client in the event of an error.
- Log files administration: Log files must be stored and maintained appropriately to avoid loss or compromise.
- Defense systems based on the signature or anomaly detection approach are not effective in protecting these applications. It is preferable to use a system based on the hybrid or machine learning/deep learning approach.
- Sensitive data should not be stored unless it is necessary.
- Validate both client- and server-side data.

2.6 Conclusion

We demonstrated in this chapter that current open-source WAFs are unsuitable for web application security. Because most of the existing WAFs are dependent on the signature detection approach. We have evaluated the effectiveness of AQTRONIX Webknight v4.4 and ModSecurity v3.0.4 in the four paranoia levels (PL) of CRS v3.3.0. The experimental results showed that Webknight is an efficient system for web application security; it has almost detected all the attacks with a recall value of 98.5%, but it has also generated a very high false positive rate with a value of 99.6%. And that ModSecurity capability depends on the PL of CRS configured, we noticed that the number of detected attacks increases with the increase of PL. Unlike other levels, ModSecurity generated a high false positive rate with a value of 60.3% at level 4.

This work opens horizons in front of several perspectives. In the future, we can propose a new generation of WAFs based on machine learning and a deep learning approach to detect both known and unknown attacks with a low false positive rate.

References

[1] R. Sturm, C. Pollard, and J. Craig, Managing Web-Based Applications, pp. 83-93, Elsevier, 2017, Available in https://doi.org/10.1016/B978-0-12-804018-8.00007-3

[2] N. K. Sangani and H. Zarger, Machine learning in application security. In : Advances in Security in Computing and Communications. IntechOpen, 2017.

[3] X. Li and Y. Xue, A Survey on Web Application Secruity, Nashville (2011).

[4] Y. Sadqi and Y. MALEH, A systematic review and taxonomy of web applications threats. Information Security Journal: A Global Perspective (2021), pp. 1-27.

[5] R. Johari and P. Sharma, A Survey On Web Application Vulnerabilities (SQLIA, XSS) Exploitation and Security Engine for SQL Injection, IEEE, 2012, Available in https://doi.org/10.1109/CSNT.2012.104

[6] H.-C. Huang, Z.-K. Zhang, H.-W. Cheng, and S. W. Shieh, Web Application Security: Threats, Countermeasures, and Pitfalls, IEEE, 2017, Available in https://doi.org/10.1109/MC.2017.183

[7] G. Betarte, R. Martınez, and A. Pardo, Web Application Attacks Detection Using Machine Learning Techniques, IEEE, 2018, Available in https://doi.org/10.1109/ICMLA.2018.00174

[8] OWASP Top 10 documentation, Available in https://owasp.org/www-project-top-ten/

[9] Y. Sadqi and M. Mekkaoui, Design Challenges and Assessment of Modern Web Applications Intrusion Detection and Prevention Systems (IDPS), Springer, 2020, Available in https://doi.org/10.1007/978-3-030-66840-2_83

[10] S. Prandl, M. Lazarescu, and P. Duc-Son, A Study of Web Application Firewall Solutions, Springer, 2015, Available in https://doi.org/10.1007/978-3-319-26961-0_29

[11] V. Clincy and H. Shahriar, Web Application Firewall: Network Security Models and Configuration, IEEE, 2018, Available in https://doi.org/10.1109/COMPSAC.2018.00144

[12] A. Moosa, Artificial Neural Network based Web Application Firewall for SQL Injection, International Journal of Computer and Information Engineering, 2010.

[13] M. Abdelhamid, Sécurité des applications Web: Analyse, Modélisation et Détection des Attaques par Apprentissage Automatique, doctoral thesis, 2016, Available in https://www.theses.fr/2016ENST0084

[14] S. Applebaum, T. Gaber and A. Ahmed, Signature-based and Machine-Learning-based Web Application Firewalls: A Short Survey, Elsevier, 2021, Available in https://doi.org/10.1016/j.procs.2021.05.105

[15] ModSecurity documentation, Available in https://github.com/SpiderLabs/ModSecurity

[16] AQTRONIX Webknight documentation, Available in https://www.iis.net/downloads/community/2016/04/aqtronix-webknight

[17] R. Ivan. ModSecurity Handbook. Feisty Duck, 2010. Available in https://www.feistyduck.com/library/modsecurity-handbook-2ed-free/online/ch01-introduction.html

[18] OWASP Core Rule Set documentation, Available in, https://owasp.org/www-project-modsecurity-core-rule-set/

[19] Payloads All The Things, Available in, https://github.com/swisskyrepo/PayloadsAll\TheThings

[20] CSIC HTTP 2010 dataset, Available in https://gitlab.fing.edu.uy/gsi/web-application-attacks-datasets

[21] Official website of bWAPP application, Available in http://www.itsecgames.com/

[22] Burp suite documentation, Available in https://portswigger.net/burp/documentation

3

Comprehensive Survey of Location Privacy and Proposed Effective Approach to Protecting the Privacy of LBS Users*

Ahmed Aloui, Samir Bourekkache, Okba Kazar, and Ezedin Barka

LINFI Laboratory, Biskra University, Algeria
College of Information Technology, UAE

Abstract

In recent years, location-based services (LBS) have become very popular, especially with the rapid emergence of the Internet of Things (IoT). However, LBS introduces new security vulnerabilities, which can lead to violations of user privacy. Therefore, protecting user location privacy has become a growing concern. In this article, we investigate the privacy attack models on LBS users. Additionally, we present an in-depth review of current protection mechanisms. Moreover, we provide the comparison between these privacy preservation mechanisms to allow new research opportunities. Furthermore, we present our approach to protecting user privacy in LBS over Euclidean space. Then, we present the comparison of our approach with other works.

Keywords: Security, location-based services, privacy preservation, attack models.

3.1 Introduction

With the rapid development of wireless technology and mobile devices, it becomes easy to locate the exact locations of users anywhere and anytime. This has made it possible to improve the applications of location-based

*User privacy in LBS.

services (LBS) very quickly. Therefore, LBS uses geographic information of mobile users based on mobile communication technologies, for example, GPS, and WLAN to provide them with information and services based on their exact locations. The applications of the mostly used LBS are geolocation advertising, geolocation marketing, geotargeting, fleet management and logistics, and traffic telematics. Thanks to LBS, users become very comfortable when requesting services. However, the use of LBS causes many problems such as great privacy breaches, accuracy of spatial information processing, and data availability [1], [2].

Preserving location privacy is essential for the successful deployment of location-based services. Currently, there are many related use case scenarios where some form of location sharing is needed and privacy should be preserved [1]. First, users send location-based queries to untrusted providers who store databases of services of interest. In this case, the privacy objective is to allow users to retrieve nearby services of interest without having to disclose exact locations to the location service provider. Moreover, public entities such as a road infrastructure monitoring network can collect large amounts of location information. This information can be in the form of instantaneous location samples (snapshots) or trajectories (continuous). In the form of trajectories, service providers collect consecutive location sequences corresponding to the same user. These datasets may be used for research or other public interest purposes [8]. In general, LBS are primarily based on the assumption that mobile users are willing to disclose their location. Additionally, in LBS, personal locations can be used as identities; so the traditional pseudonym approach does not overcome such a threat to privacy in LBS [6], [10], [11]. For example, if a user requests service closer to their personal home despite using a false identity, this will immediately result in the user's identity being revealed as a resident of the home.

Adversaries can create or launch many types of user privacy attacks when personal information such as exact locations and query contents are easily accessible by these adversaries. As a result, several types of attacks can occur, such as location attack, query attacks, and attack of homogeneity. Thereby, this leads to serious violation of user privacy when he requests service from LBS providers. This is because the LBS providers are not trustworthy because they can collect and store the personal information of the users with great spatial and temporal precision [1].

The rest of the chapter is organized as follows. Section 3.2 presents the models of privacy attack. The mechanisms of privacy protection are presented in Section 3.3. Section 3.4 presents the comparison between privacy protection mechanisms. Types of environment are presented in Section 3.5. Section 3.6 presents principles of our contributions. Section 3.7 presents our contribution in Euclidean Space. The experimentation is presented in Section 3.8. The comparison with related works are presented Section 3.9. Finally, conclusions and future works are drawn in Section 3.10.

3.2 Models of Privacy Attack

We distinguish between continuous location attack and context linking attack.

3.2.1 Continuous Location Attack

The general idea of this type of attack is that the attacker analysis a set of continuous location updates to get more information about the user, such as location or identity of the user. Certain types of attacks can occur when location continues are disclosed.

3.2.1.1 Query tracking attack

An adversary could link consecutive time snapshots to identify the originator of the query, although the location information of a query is hidden as regions [4]. For example, consider a typical query "Find the nearest gas station in the next five minutes". The service life of the query is 5 minutes. Query tracking attacks become possible if a user is hidden with different users at different times during the query duration. The idea is to exploit ***the memory property*** to defend against query tracking attacks. In the memory property, the same set of users should always be hidden together for the duration of the query [4].

3.2.1.2 Attacks of trajectory

The attacker can conclude the owner of the location when the trajectory is published, even if the identifier has been deleted. The problem of anonymizing the trajectory is to publish trajectories so that anonymity is preserved for each trajectory while maximizing the usefulness of the published data.

3.2.1.3 Identity correspondence

The adversary uses this type of attack to attack multiple user aliases. Based on attributes equal to or correlated with the same identity, the attacker links several pseudonyms. Thus, the provided privacy of the modified pseudonyms is broken [3].

3.2.1.4 Location tracking attack

This attack uses many updates of the attacker's known locations. This attack can be used against the random change of pseudonyms without using mixing zones. Moreover, even if an obfuscation mechanism is used, the attacker can correlate successive pseudonyms by linking spatial and temporal information from successive requests or position updates. For example, an attacker might try to reconstruct the movement (traffic) of users based on the available locations of multiple pseudonyms. Therefore, from intersections, the attacker can deduce where the user is or where the user's privacy-sensitive regions are. For example, the random obfuscation mechanism creates different obfuscation areas each time the user reaches their home. After that, the intersection of different obfuscation areas can be used to reduce the privacy of users.

3.2.1.5 Attack of maximum movement

In this type of attack, the attacker calculates the limited area of maximum movement when the user moves between two position updates [18]. As shown in Figure 3.1, the position of the first update executed at time T_1 helps the adversary to raise the accuracy of the update of T_2. In this example, only a small portion of the area of T_2 is accessible within the maximum movement limit. As a result, the attacker can exclude the remaining part of the position update.

The query tracking attacks are different from the attacks of the maximum movement [13]. Also, users may still suffer from maximum movement attacks, even if they are prevented from tracking requests by applying the memory property to each continuous query. Suppose two different continuous queries are issued by a mobile user at a certain time interval. The maximum movement attack shown in Figure 3.1 can still occur if the user is hidden with different sets of users for these two continuous queries. On the other hand,

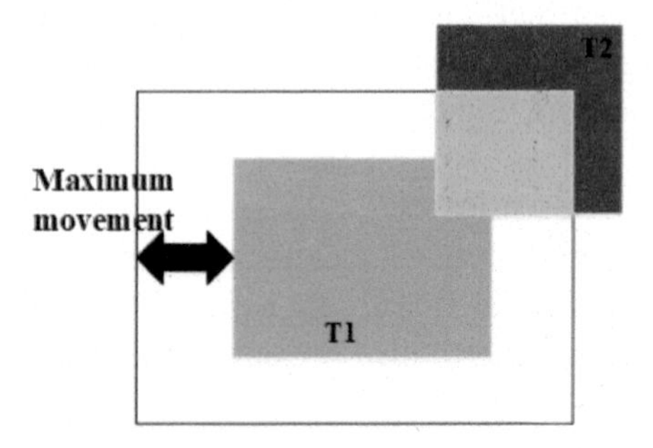

Figure 3.1 Attack of maximum movement

if the user is always hidden by the same set of users, the hidden region will extend to the entire service region when users separate and issue more and more queries over time.

3.2.2 Context Linking Attack

In this attack, contextual information is used in addition to spatial and temporal information. In order to reduce the privacy of the user, the attacker can use the knowledge of a user's personal context as well as previous knowledge, such as a map of the road networks and an address book.

In this type of attack, there are four types: attack of the personal context linking, attack of the probability distribution, attack of the map, and attack of the observation [18].

3.2.2.1 Attack of personal context linking

This attack relies on personal knowledge of the context of users, such as user preferences or interests. For example, suppose a known user regularly visits a sensitive location at a certain time and uses a simple obfuscation mechanism to protect the privacy of their location. After that, an attacker can increase his accuracy by getting a hidden position by reducing the hidden area to the location of sensitive areas [9].

3.2.2.2 Attack of observation

In this type of attack, the attacker has user knowledge gathered through observation. Therefore, an observation attack is a particular type of personal context linking attack. For example, if the attacker could see the observed user and the user is using pseudonyms, then the attacker could trace all of the user's prior locations for the same pseudonym by a single correlation [9].

3.2.2.3 Attack of probability distribution

This attack is based on environmental context information and collected traffic statistics. The attacker seeks to obtain a probability distribution function from the user's location in the hidden region. If the probability is not evenly distributed, attackers can identify areas where the user is located with a high probability [17].

3.3 Mechanisms of Privacy Protection

The privacy protection mechanism is the technique used to achieve the goal of privacy protection. We now provide an overview of existing mechanisms for protecting the privacy of LBS users. In an abstract way, the privacy protection mechanism can be defined as a transformation that maps each user's query into a different query before it becomes observable for the adversary [5]. In the most general case, the transformation consists of two functions: the first to transform the identity of the user into a pseudonym based on certain criteria and the second to transform the location [16]. To ensure that the privacy of the LBS user is protected, it is necessary to ensure that the user location is protected and the user request is protected at the same time. That is why there are two types of privacy protection. The first type is called location privacy, which requires adversaries (unauthorized parties) not to be aware of the user's current or past location. The second type is called query privacy, which requires that adversaries must not be able to tie a specific request to a specific user. There are a number of works representing the state of the art mechanisms to guarantee the privacy of the user. In this section, we will present an overview of the basic principles of these mechanisms.

3.3.1 Cloaking

This mechanism uses the k-anonymity model to ensure location privacy. The k-anonymity model is concept not limited to security. k-anonymity ensures that the adversary cannot distinguish the sender of the request from $k - 1$ other users [9].

Therefore, the basic idea behind this mechanism is that a user sends a region to service providers instead of their exact location. Additionally, this region must contain the exact location of the query submitter and the locations of at least $k - 1$ other users. The cloaking mechanism ensures that the service provider will only receive a certain interval, and never receive accurate information.

3.3.2 Cryptography

The technique of encryption (such as symmetric encryption) is more used in cryptographic approaches to ensure the privacy of the location. In [12], the authors use symmetric encryption to ensure hiding the user's current location from the LBS provider.

On the other hand, in [7], the authors use personal information retrieval (PIR) technique to ensure the privacy of the location.

3.3.3 Obfuscation

These mechanisms use masking techniques. For example, they use geo-coordinates conversion to hide users' coordinates, which allows a wider geographic area rather than the exact location to be sent to service providers. Therefore, the main objective of this mechanism is to deliberately decrease the quality of information in order to ensure the protection of user privacy. In addition, these mechanisms use the k-anonymity model in order to evaluate the protection of location privacy. The main idea behind these mechanisms is to use the user's location history to create hidden regions [14], [15].

In other words, this mechanism requires the use of a larger area which must necessarily contain the actual location of the user. Therefore, we can define location obfuscation as "is a means of deliberately degrading the quality of a user's location information" [6].

3.3.4 Dummies

Dummy queries are used to provide anonymization of a location. In this protection technique, the exact location of the user is transmitted together with a group of dummies (other fictitious locations) to the service provider. Therefore, the main purpose of using Dummies is to keep a user's exact location private. In this case, service providers are obligated to resend responses for each Dummy location. In addition, these mechanisms use the k-anonymity model in order to evaluate the protection of location privacy.

Additionally, this mechanism ensures that the use of a third-party TTP can be dropped. In other words, the user can create Dummies without having to use a third-party TTP.

However, if the adversary has contextual information, he can easily distinguish between dummy requests and real requests. For example, if a Dummy's location is in a lake, then the opponent can easily drop this Dummy.

3.3.5 Mix-zones

In the mix-zone mechanism, user locations are hidden in mix zones. The advantage of this mechanism, is that it ensures that the locations of users are not known in the mixing zones. Additionally, in these mixing zones, users do not send location updates.

Moreover, to ensure the success of this mechanism, it is necessary that each user has at least one mixing zone. In this way, user locations cannot be retrieved by service providers [3].

Therefore, the basic idea of this mechanism is to change the identities of users of the zone among themselves. In this way, the attacker cannot link the different user identities.

3.4 Comparison between Privacy Protection Mechanisms

It is necessary to have a summary or a good synthesis of many existing mechanisms to guarantee the best functionalities for our needs. Table 3.1 presents a good summary of many mechanisms.

In order to provide at least an initial tool to compare protection mechanisms, some key factors are proposed [1]:

- Needs a TTP (trusted third party)
- Reports any information of location to the LBS provider
- Needs special implementation on the LBS provider side
- Effectiveness (privacy and utility)
- Query types: snapshots or continues

Figure 3.2 presents the tradeoff between privacy and the performance of protection approaches [1].

Cryptographic methods provide complete privacy protection with no location information being disclosed. However, in terms of performance,

Table 3.1 Comparison of different mechanisms for preserving location privacy

Mechanisms	TTP	Location information	Snapshots/ continue	Privacy	Utility	Special implementation
Obfuscation	No	Region	S/C	Yes	Yes	No
Dummies	No	Yes	S	Yes	No	No
Mix-zones	Yes	Yes	S	Yes	No	No
K-anonymity	No	Region	S/C	Yes	Yes	No
Spatial cloaking	No	Region	S/C	Yes	Yes	No
Cryptography	No	No	S/C	Yes	No	Yes

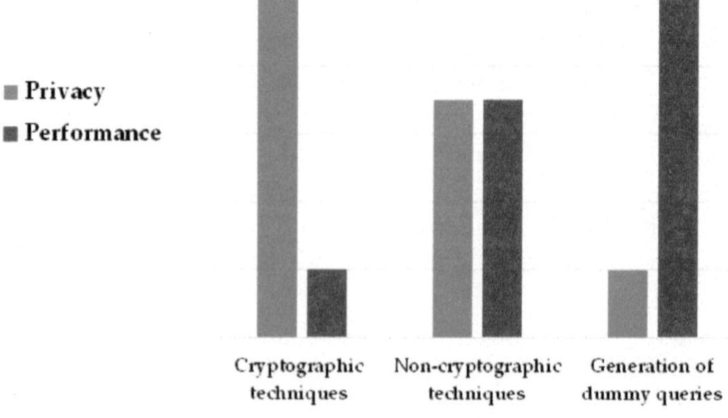

Figure 3.2 Tradeoff between privacy and performance

cryptographic operations can result in higher request processing costs and the costs of calculation and communication on the server.

Non-cryptographic methods such as spatial transformation approaches and cloaking approaches provide a certain level of protection at low cost while preserving the geographic validity of data and its availability for use. Additionally, these approaches result in request processing overhead, where the region request is processed in a cloaking region, as shown in Figure 3.2.

3.5 Types of Environment

In the literature, there are researches that have been proposed to protect the privacy of users in Euclidean space and others in road networks.

In road networks, the movement of users is limited by segments of the road networks. Movement limits such as the user's maximum speed limit and there are also the directions of road network segments. On the other hand, there are no limits to the movements of users in Euclidean space. The user can move freely in Euclidean space, i.e., there are no direction or speed constraints.

Figure 3.3 represents the difference between Euclidean space and road networks. Figure 3.3(a) represents the movement of the user in Euclidean space. As shown in this Figure 3.3(a), the user can cross directly between two points i and j without respecting the segments of the road network.

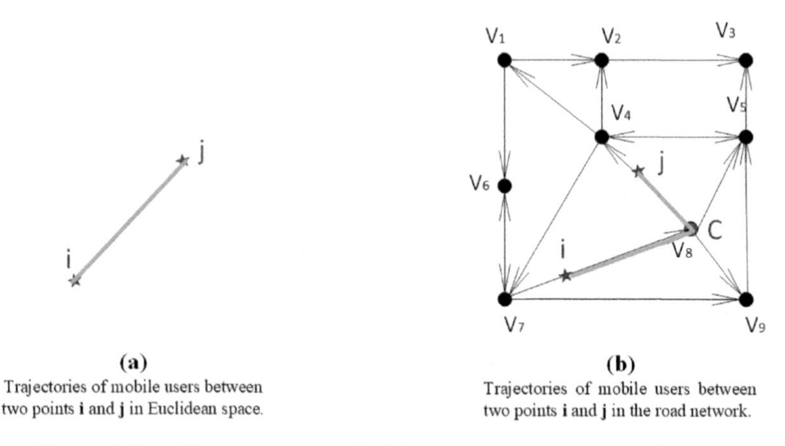

Figure 3.3 Difference between Euclidean space and road network space

Figure 3.3(b) represents the movement of the user in road networks. Figure 3.3(b) represents user movement in road networks. The user must pass through point C in order to cross between two points i and j of the road network

3.6 Principles of Our Contributions

In this chapter, we propose an approach to ensure the protection of user privacy. This approach is proposed to ensure user privacy protection in Euclidean space.

Compared to other approaches, our approach allows users to determine parameters of privacy requirements, such as minimum requirements at different locations.

Moreover, in our approach, we considered speed similarity and direction similarity to ensure that users have a higher probability of staying together in the future.

In our approach of Euclidean space, we considered hiding the user's location history to ensure continued query protection.

The k-anonymity model has been considered a protection requirement in our approach. The main advantage of this model is that it makes each user indistinguishable from $k - 1$ other users. Thus, our approach extends the exact location of the user to a hidden region for the Euclidean space approach. In addition, k_{global} requires that the intersection of current sets (regions) with previously created sets be greater than k_{global}. This condition is used in continuous queries.

Furthermore, we view A_{min} and A_{max} as a privacy requirement to guarantee a better quality of services QoS. The A_{max} and A_{min} parameters specify the maximum and minimum lengths that a masked region must have. The user will feel uncomfortable revealing the small hidden region because any point within this small region will be closer to the user's exact location. It is for this reason that we used the parameter A_{min}. Moreover, the QoS is bad in the case where the value of A_{max} is higher. This is because the exact location of the user may be far from the service.

In our approach, we used the Clique principle. The main advantage of using the clique is that it is very effective against location attacks.

In order to prevent continuous location attacks, our approach is based on the MMB and MAB methods. The protection principle of the MAB method is that it requires that the hidden region at time t_i be completely covered by the user's MAB at time t_{i+1}. The principle behind the protection of the MMB method is that it also requires that the hidden region at time t_{i+1} be completely covered by the MMB of the user at time t_i.

For continuous queries, if the condition of k_{global} is met, our approach continues to send the set of hidden region to the LBS providers.

To deal with query attacks in our approach, we considered the motion similarity of users. This is ensured by calculating the direction similarity and the velocity similarity. It is for this reason that the (k) users should have similar moves. Additionally, the requestor must be masked with at least $k-1$ other users. Also, these k users must be in the form of a clique. These last constraints are very important to ensure overall confidentiality.

3.7 Our Contribution in Euclidean Space ES

In our approach, to deal with these attacks, we model the user privacy requirements by an undirected graph $G(V, E)$ where the set V represents the user locations, and the set E is a set of edges. As shown in Figure 3.4(a); each user has their own MMB.

We used the MMB distance to determine neighbors and build edges between mobile users. There is an edge e between two users MC_1 and MC_2, if and only if:

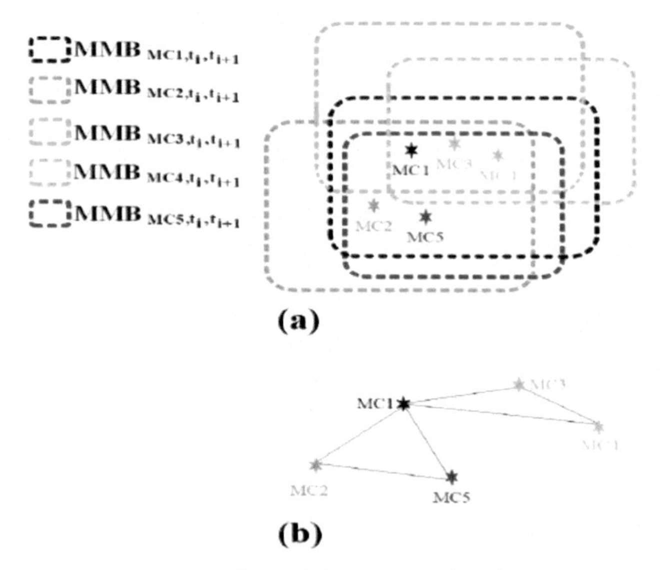

Figure 3.4 Graphe $G(V, E)$

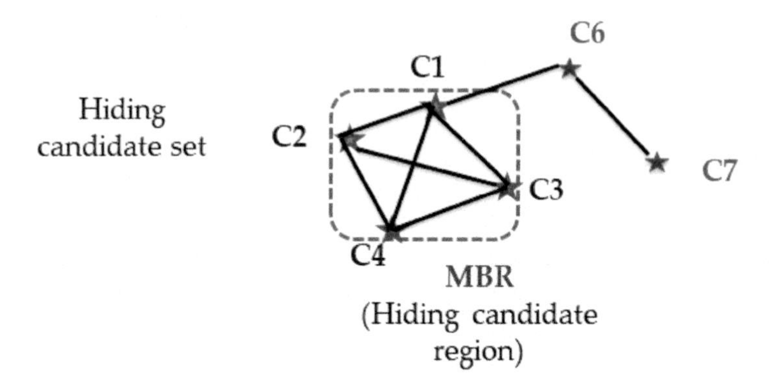

Figure 3.5 Clique $C = (S, A)$

- the user MC_1 is covered by the MMB of the user MC_2;
- the user MC_2 is covered by the MMB of the user MC_1.

After the construction of the graph, we selected the cliques. A clique $C = (S, A)$ of a graph $G = (V, E)$ is a complete subgraph, i.e., any two vertices of the clique are always adjacent.

As shown in Figure 3.5, the four users C_1, C_2, C_3, and C_4 form a clique. All of these clique users represent a hiding candidate set and the MBR of their locations is used as the candidate hiding region. We used the clique to prevent the adversary from excluding users from the hiding set and to ensure that users stay together.

3.7.1 Method of Selection of Hiding Candidate Set in ES

In this contribution, we have proposed a method to select the candidates UC. A mobile user set is a cloaking candidate set if and only if:

- Clique: the set of users must be in the form of a clique.
- Speed and direction similarity: the set of users must have the same similarity of movements.
- Location anonymity $k <= UC$: the number of users must be greater than k.
- Cloaking granularity $A_{\text{min}} <= MBR(EC) <= A_{\text{max}}$: the MBR of the set of users must meet the constraints of minimum and maximum areas.
- L-diversity: the set of users must have at least L different queries to deal with the homogeneity attack.

After fulfilling all these conditions, the MBR (minimum bounding rectangle) of this set is a candidate region for hiding.

3.7.2 Method of Creating Qualified Hiding Region

Then, if this candidate hiding region CRt_i fulfills the following conditions, it is then a qualified hiding region R_{t_i}:

- Properties of MMB and MAB: The distance between the candidate region and the previous region at time t_{i-1} must respect the properties of MMB and MAB to cope with the location attack.
- Global anonymization: The size of the intersection of this candidate region of hiding CR$_{ti}$ with those created previously must be greater than K_{global} to ensure global anonymization.
- K-sharing property: This candidate region must satisfy the k-sharing property to deal with the linking attack.

After having fulfilled all these conditions, this candidate region is a qualified hiding region which is transmitted to the provider.

3.7.3 Operation of Our Approach

As shown in Figure 3.6, in our approach, the mobile user sends a request to a central server, where the request is processed and hidden with other clients. Then, the central server builds the region request and then it transmits this region request to the suppliers instead of the user request. Then the provider sends all results to the central server. After that, the central server filters the

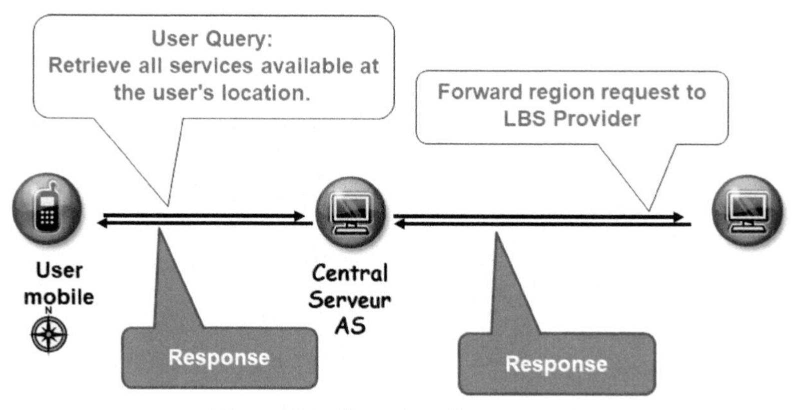

Figure 3.6 Operation of our approach

results and sends them to users based on their exact location. The provider is a critical place where customer privacy can be violated by location context and query content.

3.7.4 Hiding Principle of Our Approach

The basic idea of our protection approach is to find a set of hiding that satisfies the user privacy requirements and the conditions of our methods. We can divide our approach into four steps, as shown in Figure 3.7:

- Receipt of queries: After receiving the request, the query is passed to the protection process to generate a hidden region according to the privacy requirements of the users.
- Graph and cliques: We build the MMB for each user to determine the neighbors and build the graph. Our hiding method is clique-based in which the central server will continue to search for cliques that contain the query submitted.
- Hiding candidate set: Then, we used the two methods that we have proposed to select the candidate sets and create the qualified hidden region.

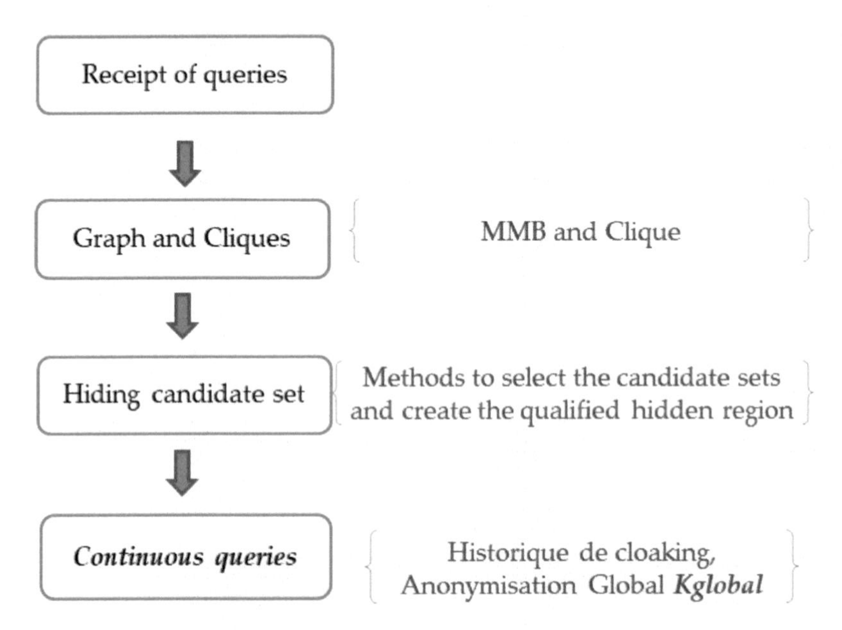

Figure 3.7 Hiding principle

- Continuous queries: For continuous queries, our approach is history-based, i.e., to create the hidden region at the next snapshot, the central server uses users from previously hidden regions. In addition, during the lifetime of the continuous query, the central server continues to send the region request to LBS provider if the K_{global} condition is met.

3.7.5 Generate Dummies (Dummy Queries)

In our approach, one creates dummy queries because the other core approaches will underperform if the k-anonymity requirement is not met. So, in this case, central approaches have two options:

- Wait for $k - 1$ other queries to be made in the same region.
- Expand the region looking for $k - 1$ other queries until the region contains more than $k - 1$ other clients.

But the problem of these central approaches, if the central approach still cannot meet the k-anonymity requirement after these two options, the issuer's request will be deleted. That is why we created dummies in our approaches.

For this, to ensure the reliability of our work, we generated Dummies instead of removing the user request to meet the k-anonymity requirement.

As shown in Figure 3.8, there are only three users in the region and the value of $k = 4$; so we create a Dummy to meet the k-anonymity requirement.

The opponents can easily find the difference between the real user and the Dummies in case these Dummies are randomly generated. Therefore, in our approach, the Dummies must satisfy the temporal and spatial properties of the real user query (which look like real users).

Figure 3.8 Generate Dummies

3.8 Experimentation

We used the mobile user generator (Brinkhoff) in order to test the effectiveness of our approach. The input to the generator is an actual city road network.

We used three parameters to evaluate our approach: we start with guarantee of privacy which is represented by K_p. The guarantee of privacy represented by the K_p:

$K_p = (R_0 \cap R_1 \cap R_2 \cap R_n / |R_0|) * 100.$

As shown in Figure 3.9, compared with D-TC and V-DCA, our approach has a relatively larger K_p when the number of snapshots reaches 90 snapshots for V-DCA and 70 snapshots for D-TC. Furthermore, our approach is much more stable, with a 15% for more than 120 snapshots.

Therefore, the results indicate that our approach keeps more clients from the first clocking region (R_0) in consecutive snapshots. This is thanks to our method of selecting the hiding set.

The second metric is QoS. As shown in Figure 3.10, compared with D-TC and V-DCA, the average hiding area and average distance of our approach are

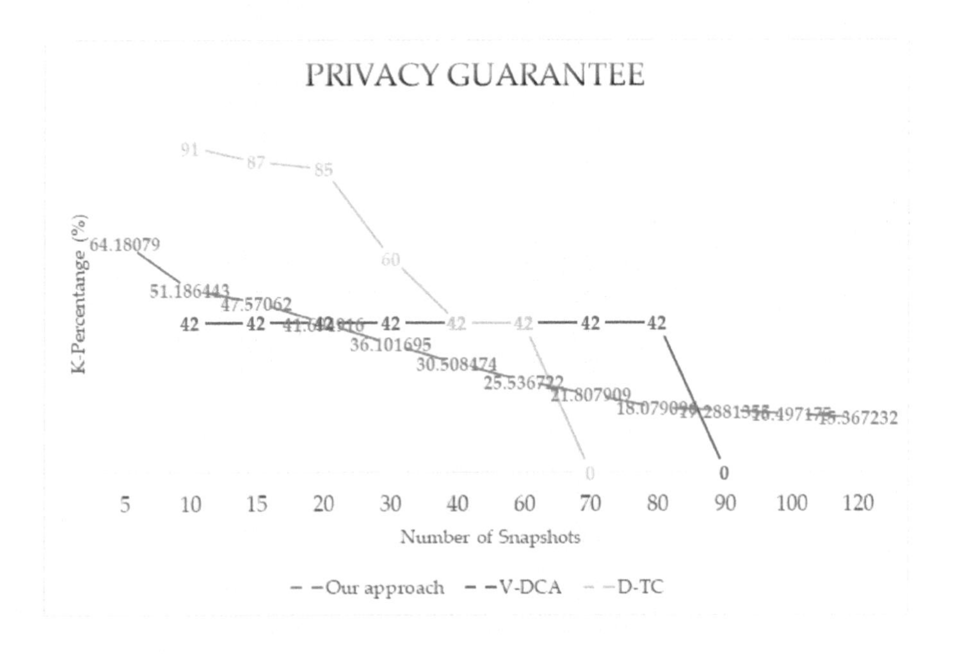

Figure 3.9 Guarantee of privacy

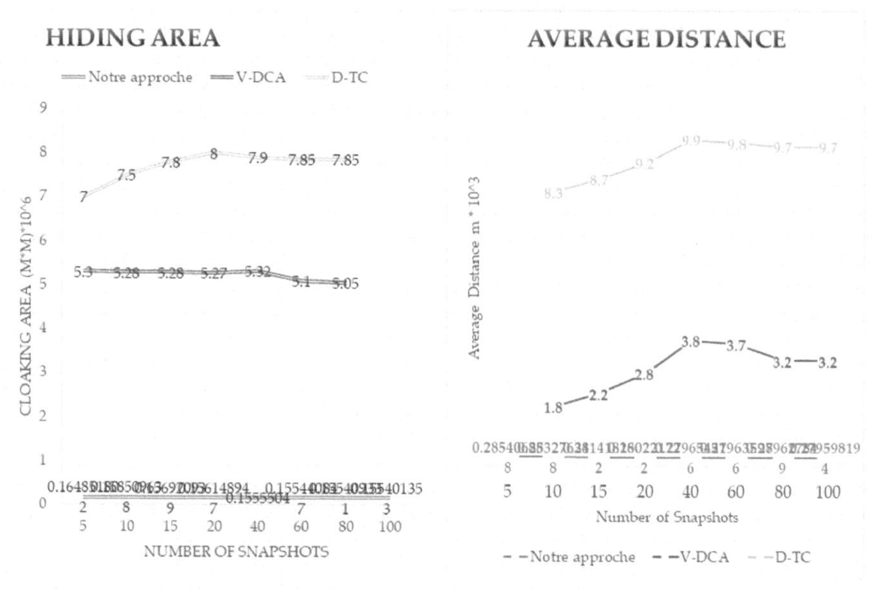

Figure 3.10 Quality of service QoS

much smaller and more stable than other approaches, which shows that our approach offers (delivers) more stable QoS and higher than other approaches.

In addition, according to these results, and previous results it is indicated that the other approaches preserve privacy, but this leads to lower QoS. So they do not provide a good balance between privacy and QoS.

The third metric is performance, which represents the time it takes to find the cloaking region. As shown in Figure 3.11, in the first snapshots, our approach performs better than the GCA approach. But the other V-DCA and D-TC approaches work better than our approach. Since, in our approach, we considered more factors to ensure a good balance between privacy and QoS, such as clique building, motion similarity, A_{min} and A_{max}. In addition, the process of discovering the cliques takes a lot of time. In addition, we have taken into account several types of attack in our approach.

3.9 Comparison with Related Works

As shown in Figure 3.12, there are approaches that have adopted one or two privacy measures. There are approaches that are only interested in location protection, and others by query privacy protection.

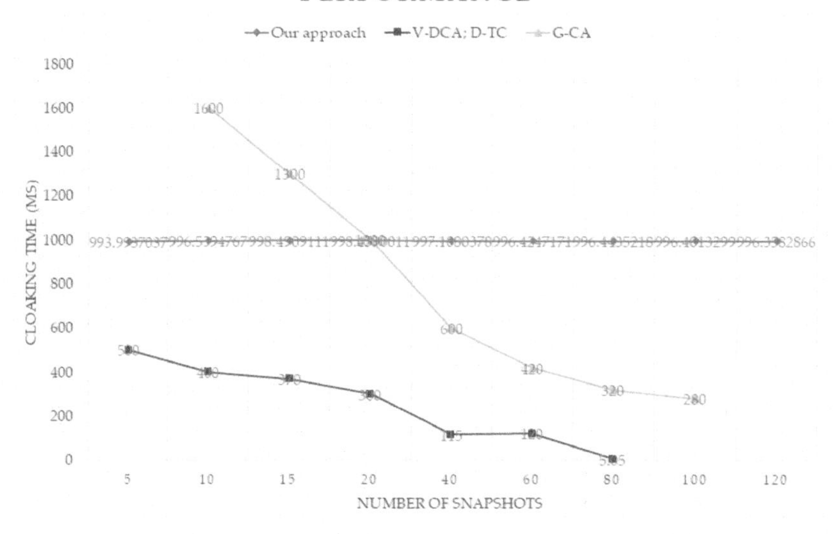

Figure 3.11 Performance

	Global of privacy			The attacks				Static/ Dynamic	Privacy protection Types		Privacy Metrics
	Loc	Id	Time	Location	Linking	Query sampling	Homog	Snap/ Cont	Query Privacy	Location Privacy	
Pan et al 2012	Yes	Yes	Yes	Yes	Yes	No	No	C	No	Yes	2
Memon et al 2015	No	Yes	Yes	No	Yes	No	Yes	C	Yes	No	2
Saravanan et al 2016	Yes	Yes	Yes	Yes	Yes	No	No	S/C	No	Yes	2
UM et al 2009	No	No	Yes	No	No	No	No	S	No	Yes	2
Bambaet al 2008	No	No	Yes	No	No	No	No	S	No	Yes	2
Lee et al 2012	No	No	Yes	No	No	No	No	C	Yes	No	1
Wang et al 2012	No	Yes	Yes	No	Yes	Yes	No	C	Yes	No	1
Our Approach	**Yes**	**Yes**	**Yes**	**Yes**	**Yes**	**Yes**	**Yes**	**S/C**	**Yes**	**Yes**	**3**

Figure 3.12 Comparison with related works

Moreover, the other approaches provided special solutions against a single type of attack. On the other hand, in our approach, we considered several types of attacks.

3.10 Conclusion

In this chapter, several models of user privacy attack have been presented. Moreover, we presented a review of current protection mechanisms. Privacy attacks are a critical threat because they threaten the privacy of LBS users. It is very important to implement protocols and security mechanisms to control the entities of the LBS, preserve the security of users, and also allow users to use the LBS system with confidence. Moreover, we have presented an approach to ensure the protection of user privacy in LBS over Euclidean space.

We cannot immediately apply Euclidean spatial mechanisms to the road network environment because this mechanism leads to inefficient query processing and loss of privacy. This is why we will propose another approach to ensure the protection of the privacy of users of LBS designed for road networks.

Acknowledgment

This research was partially supported by Intelligent Computer Science Laboratory LINFI of Algeria. We thank some colleagues from Laboratory LINFI who provided insight and expertise that greatly assisted the research.

References

[1] Ahmed Aloui and Okba Kazar. A survey on privacy preservation in location-based mobile business: Research directions. *International Journal of Web Portals (IJWP)*, 13(1):20–39, 2021.

[2] Ahmed Aloui, Okba Kazar, Samir Bourekkache, and Fouzia Omary. An efficient approach for privacy-preserving of the client's location and query in m-business supplying lbs services. *International Journal of Wireless Information Networks*, 1–22, 2020.

[3] Alastair R. Beresford and Frank Stajano. Mix zones: User privacy in location-aware services. In *Proceedings of the 2d IEEE Annual Conference on Pervasive Computing and Communications Workshops*, pages 127–131. IEEE, 2004.

[4] Chi-Yin Chow and Mohamed F. Mokbel. Enabling private continuous queries for revealed user locations. In *International Symposium on Spatial and Temporal Databases*, pages 258–275. Springer, 2007.

[5] Maria Luisa Damiani. Location privacy models in mobile applications: conceptual view and research directions. *GeoInformatica*, 18(4):819–842, 2014.

[6] Matt Duckham and Lars Kulik. A formal model of obfuscation and negotiation for location privacy. In *International Conference on Pervasive Computing*, pages 152–170. Springer, 2005.

[7] Gabriel Ghinita, Maria Luisa Damiani, Claudio Silvestri, and Elisa Bertino. Preventing velocity-based linkage attacks in location-aware applications. In *Proceedings of the 17th ACM SIGSPATIAL International Conference on Advances in Geographic Information Systems*, pages 246–255. ACM, 2009.

[8] Liliana Gonzalez, Pedro Wightman Rojas, M Labrador, *et al.* A survey on privacy in location-based services. *Ingeniería y Desarrollo*, 32(2):314–343, 2014.

[9] Marco Gruteser and Dirk Grunwald. Anonymous usage of location-based services through spatial and temporal cloaking. In *Proceedings of the International Conference on Mobile Systems, Applications and Services*, pages 31–42. ACM, 2003.

[10] Ali Khoshgozaran, Houtan Shirani-Mehr, and Cyrus Shahabi. Spiral: A scalable private information retrieval approach to location privacy. In *2008 9th International Conference on Mobile Data Management Workshops, MDMW*, pages 55–62. IEEE, 2008.

[11] Eyal Kushilevitz and Rafail Ostrovsky. Replication is not needed: Single database, computationally-private information retrieval. In *Proceedings 38th Annual Symposium on Foundations of Computer Science*, pages 364–373. IEEE, 1997.

[12] Sergio Mascetti, Dario Freni, Claudio Bettini, X. Sean Wang, and Sushil Jajodia. Privacy in geo-social networks: Proximity notification with untrusted service providers and curious buddies. *The VLDB Journal-The International Journal on Very Large Data Bases*, 20(4):541–566, 2011.

[13] Xiao Pan, Jianliang Xu, and Xiaofeng Meng. Protecting location privacy against location-dependent attacks in mobile services. *IEEE Transactions on Knowledge and Data Engineering*, 24(8):1506–1519, 2012.

[14] Aniket Pingley, Nan Zhang, Xinwen Fu, Hyeong-Ah Choi, Suresh Subramaniam, and Wei Zhao. Protection of query privacy for continuous location based services. In *2011 Proceedings IEEE INFOCOM*, pages 1710–1718. IEEE, 2011.

[15] Daniele Quercia, Ilias Leontiadis, Liam McNamara, Cecilia Mascolo, and Jon Crowcroft. Spotme if you can: Randomized responses for location obfuscation on mobile phones. In *2011 31st International Conference on Distributed Computing Systems*, pages 363–372. IEEE, 2011.

[16] Reza Shokri, Panos Papadimitratos, George Theodorakopoulos, and Jean-Pierre Hubaux. Collaborative location privacy. In *2011 IEEE 8th International Conference on Mobile Ad-Hoc and Sensor Systems*, pages 500–509. IEEE, 2011.

[17] Reza Shokri, George Theodorakopoulos, Jean-Yves Le Boudec, and Jean-Pierre Hubaux. Quantifying location privacy. In *2011 IEEE Symposium on Security and Privacy,* pages 247–262. IEEE, 2011.

[18] Marius Wernke, Pavel Skvortsov, Frank Dürr, and Kurt Rothermel. A classification of location privacy attacks and approaches. *Personal and Ubiquitous Computing*, 18(1):163–175, 2014.

4

Analysis of Encrypted Network Traffic using Machine Learning Models

Aradhita Bhandari, Aswani Kumar Cherukuri, and Sumaiya Thaseen Ikram

School of Information Technology and Engineering,
Vellore Institute of Technology, India
E-mail: aradhita.bhandari2018@vitstudent.ac.in; aswani@vit.ac.in;
sumaiyathaseen@gmail.com

Abstract

Traffic classification is essential for identifying normal behavior from diverse traffic. Modern applications that contribute to half of the world's traffic generate encrypted traffic to protect the data between clients and servers. Encryption is required to prevent adversaries from accessing private information such as user behavior patterns, application data, and passwords. Ideally, this encrypted traffic would be legitimate; however, in some cases, the encrypted traffic can conceal malicious software like viruses, worms, and Trojans. Therefore, to secure our networks, large-scale encrypted traffic analysis is required in real time to identify abnormal patterns as soon as possible. This chapter will discuss various unsupervised, supervised, and semi-supervised learning techniques deployed to analyze encrypted traffic. This paper presents the first such analysis for encrypted network data to the best of our knowledge. The learning models analyzed are K-means clustering, random forest, label propagation, and AdaBoost classifiers. Experimental analysis is performed on UNSW-NB15 encrypted traffic datasets.

Keywords: Abnormal pattern, clustering, encrypted traffic, random forest, supervised learning, unsupervised learning.

4.1 Introduction

Network traffic has to be analyzed to diagnose problems, provision resources, and identify the network's attacks or misuse. Malicious packets are not rare and often occur after an intruder gains access to the network or when legitimate users attempt to interrupt network policies. Traffic characterization is problematic due to the increased use of cryptographic protocols hindering traditional traffic analysis.

Many researchers address the classification problem of encrypted network traffic by proposing various solutions. For example, an encrypted web traffic classification scheme [1] is developed by considering only timing information. Encrypted webpage traffic [2] is identified based on the bidirectional communication between clients by analyzing only the packet length information. Barut *et al.* [3] comprehensively investigated a collection of commonly used machine learning and deep learning techniques to detect encrypted malware. Detailed tests are conducted to measure the throughput and CPU and RAM consumption. In addition, the classification accuracy of the algorithms is also determined. Leroux *et al.* [6] analyzed using features derived from encrypted streams. The features are trained using machine learning approaches to anticipate the traffic type passing through TOR or IPsec tunnel. They employ compact machine learning models that operate on minimal window size. As a result, this technique can be used in real time. The common problem that is unsolved is that the performance of each technique is limited to the specific dataset.

Several challenges are to be addressed to achieve this goal. Attribute collection for encrypted traffic is the first challenge essential for traffic classification by exploring the data. The second challenge is identifying a machine learning classifier that can result in optimal performance. Critically, real-world data for attack analysis is often highly skewed, affecting the performance of classifier models. Further, labeling a large amount of data is often not possible, especially in the case of encrypted traffic analysis. Thus, there arises the need for a highly efficient semi-supervised machine learning method that can handle naturally skewed data.

This chapter will address the two challenges by deploying efficient feature selection and machine learning techniques.

The objectives of this research work are as follows:

i) Developing a highly efficient technique using machine learning to identify malicious activity in encrypted network traffic.

ii) Determining the effectiveness of novel approach of unsupervised algorithms for encrypted network traffic fingerprinting.
iii) Performing a comparative analysis between the chosen unsupervised and supervised learning algorithms for encrypted network traffic analysis.
iv) Using various techniques to improve the results of the two best-performing networks.

4.2 Literature Review

Significant research has been conducted on various machine learning classifiers for traffic analysis. Limthong and Tawsook [7] utilized two well-known machine learning algorithms: Naive Bayes and *k*-nearest neighbor. The authors of this paper conducted experiments to investigate the association between time-based network traffic variables and various forms of network abnormalities. Their findings will aid researchers and network managers in selecting appropriate time-based characteristics for each sort of anomaly, as well as choosing a machine learning method for their system.

Akbari *et al.* [10] developed a neural network approach, and they attained a service classification accuracy of 95% with minimal parameters. Cai *et al.* [4] conducted a systematic analysis of attacks and defenses to determine which traffic features convey the most information; established lower bounds on the bandwidth costs of any defense that achieves a given level of security; and presented a mathematical framework for evaluating fingerprinting attacks and defenses in the open world based on their closed-world performance. Cherubin and Juarez [5] suggested two WF defenses Tor. Onion sites, one for the server and one for the client, work at the application layer. The advantages of this layer are: (i) It allows for granular control of webpage content, which is probably why WF attacks are viable, and (ii) it makes defenses simple to construct.

Moustafa and Slay [8] created a UNSW-NB15 dataset, a fusion of legitimate and attack traffic generated in the campus of UNSW, and analyzed existing and novel techniques for traffic identification. In addition, Moustafa and Slay [9] demonstrated the complexity of the UNSWNB15 dataset. The correlations and statistical analysis of the attributes were examined. Five existing classifiers were used to analyze the performance, and this dataset is one of the benchmark datasets for evaluating NIDSs.

Perera *et al.* [11] compared the results of six commonly used supervised machine learning algorithms, and it is observed that the random forest and decision tree techniques produce superior performance. Velan *et al.* [12] analyzed encrypted traffic and its classification. The authors identified the encrypted application protocol and the encryption protocol. Cherukuri *et al.* [13] studied the various flow-based approaches for traffic categorization and discussed the mechanisms to implement integrity of flows between senders to collect in an IoT network. Jha *et al.* [14] deployed various machine learning methods to identify the DNS over HTTPS (DoH) tunneling according to the request duration and packet size. Ikram *et al.* [15] utilized different neural network models to develop a robust anomaly detection approach on the UNSW_NB and custom generated datasets. An ensemble classifier XGBoost assimilates the results of various neural network models to result in higher accuracy.

Therefore, in this chapter, an extensive literature of various machine learning models is studied, and experiments are performed using encrypted traffic datasets to analyze the performance of the classifiers.

4.3 Background

4.3.1 Supervised Learning

4.3.1.1 AdaBoost

Adaptive boosting is a machine learning approach created to improve the results of traditional learning methods using simple decision trees. It consists of multiple single-split decision trees called "weak learners." These weak learners then work together until they can be combined into a single "strong" classifier.

The algorithm "boosts" its learning process by focusing on incorrect predictions more than that on the predictions made correctly, hence the name Adaptive. By doing so, the algorithm can focus on correcting its mistakes rather than expending energy on already correct predictions.

The model works in the following steps:

Step 1: A sequence of single-split decision trees (weak learners) is trained on the data using arbitrary and equal initial weights.

Step 2: An initial prediction is found and compared to the true labels.

Step 3: The incorrect predictions are given a higher weight than others.

Step 4: The process is repeated iteratively with more learners until the model reaches a limit in the number of learners or accuracy.

4.3.1.2 Random forest

Random forest is a well-known supervised machine learning approach. It is an ensemble model that integrates several classifiers and increases its performance. The classifier contains a number of decision trees on various subsets of a given dataset and takes the average to enhance the predicted accuracy of that dataset.

4.3.2 Unsupervised Learning

4.3.2.1 K-Means clustering

K-means clustering is an unsupervised learning technique that divides the unlabeled dataset into various clusters. In this technique, "K" specifies the number of predefined clusters that must be produced during the process; for example, if $K = 2$, two clusters will be produced, and if $K = 3$, three clusters will be produced. In this centroid approach, each cluster has its unique centroid. The primary goal of this technique is to reduce the sum of distances between data points and the clusters that they belong to.

The technique takes an unlabeled dataset as input, separates it into a K-number of clusters, and continues the procedure until no better clusters are found. In this algorithm, the value of K should be predetermined.

The steps for implementing the K-means algorithm are as follows:

Step 1: To determine the number of clusters, choose the K.
Step 2: Choose K locations or centroids at random. (It could be something different from the incoming dataset.)
Step 3: Assign each data point to the centroid closest to it, forming the preset K clusters.
Step 4: Determine the variance and reposition each cluster's centroid.
Step 5: Reverse the third step, reassigning each datapoint to the cluster's new nearest centroid.
Step 6: If there is a reassignment, go to step 4; otherwise, move to FINISH.

4.3.3 Semi-Supervised Learning

4.3.3.1 Label propagation

Label propagation is a fast algorithm that detects communities in a graph, similar to how a clustering algorithm works. It relies on the logic that labels will "propagate" quickly amongst densely connected nodes but take a long time to travel a large distance to a sparsely connected node. Hence, similar nodes will end up with the same label. Since the model expects partially labeled data, it uses random unique label values and discards the rest.

The algorithm works in the following steps:

Step 1: Every node is initialized with a label.
Step 2: The labels propagate within the network through iterations in which each node updates its label to match the majority of its neighbors. Ties are broken arbitrarily.
Step 3: This proves repeats until user-defined maximum is reached or convergence of labels occurs with no new learning.

4.4 Experimental Analysis

4.4.1 Dataset

The raw network traffic is generated in the cyber range lab of UNSW Canberra for developing a fusion of real and synthetic contemporary attack behaviors. The training set used for this paper had 44 different features and contained data about 175,341 rows, where each row represented an encrypted packet.

4.4.2 Feature Analysis

The *attack_cat* feature of the dataset denotes the attack label. The column is described in Table 4.1. Table 4.1 also includes the results of label encoding the *attack_cat* column.

It is evident from Table 4.1 that the target feature is heavily imbalanced, with "Worms" making up merely 0.07% of the entire dataset. Non-attack data consists of more than 30% of the whole data and the top three attack types constitute more than 50% of the whole data. To draw inferences on the data, the correlation between the various features is calculated to show the relationship between the data. Pearson's correlation coefficient indicates the linear association among the data; if the relationship between the data can be represented perfectly linearly, the magnitude of correlation will be 1 and positive for positively correlated and negative for negatively correlated. Uncorrelated values will have a magnitude of 0. The correlation is calculated using the pandas' method DataFrame.corr(). It is observed that a considerable number of the columns are not correlated to the target class. This can lead to a reduction in the accuracy of the learning. Further, many non-target features

Table 4.1 The description of labels in the *attack_cat* column

Label	Description	Count	Percentage	Encoded
Normal	Benign packet	56,000	31.94%	6
Generic	Attack against block ciphers	40,000	22.81%	5
Exploits	Exploitation of a vulnerability in system	33,393	19.04%	3
Fuzzers	Feeding randomly generated data to program or network to disrupt functioning	18,184	10.3%	4
DoS	Services of a host connected to the Internet are suspended	12,264	6.99%	2
Reconnaissance	Information gathering attacks	10,491	5.98%	7
Analysis	Such as portscan, spam, and html file penetrations	2000	1.14%	0
Backdoor	Bypassing of security mechanisms to access a computer or its data	1746	1.00%	1
Shellcode	Code payload to exploit vulnerability	1133	0.65%	8
Worms	Self-replicating memory corruptors	130	0.07%	9

are highly correlated, which means that they do not contribute much to the training but consume more resources and may reduce the accuracy.

First, the data that correlates less than 0.1 in magnitude to the target class is removed. Only 20 features are retained, as shown in Table 4.2 It is important to note that the highest correlation magnitude between another attribute and the target class is 0.3.

Then, each feature was compared to the other features except for the target label for correlation magnitude higher than 0.65. This resulted in nine features for training (*proto*, *rate*, *sttl*, *sload*, *dload*, *smean*, *dmean*, *ct_flw_http_mthd*, and *is_sm_ips_ports*). All such features were dropped.

A lower number of features significantly reduce the resource overhead for training and use. Since this is meant to be a real-world system, the lightness of the model and ease of retraining were considered during the feature selection process. More features can be added if sufficient accuracy is not achieved.

Table 4.2 Features extracted after Pearson's correlation

Feature	Correlation value
dload	0.18287
dmean	0.228046
proto	0.729044
ackdat	0.229887
is_sm_ips_ports	0.238225
dttl	0.279989
ct_flw_http_mthd	0.290195
dpkts	0.292931
smean	0.303855
sload	0.305579
ct_dst_ltm	0.318517
ct_src_ltm	0.328629
ct_dst_src_ltm	0.337979
ct_srv_src	0.341513
ct_srv_dst	0.357213
ct_state_ttl	0.393668
rate	0.504159
ct_src_dport_ltm	0.577704
ct_dst_sport_ltm	0.692741
sttl	0.727173

4.4.3 Pre-Processing

Since only nine features are used for training, it is important to remove any issues that may cause a reduction in the quality of training. High skewness can cause the model to misconstrue all the data in the "tail" region as outlier data, negatively affecting model performance. The skewness of each of the columns is shown in Table 4.3 As the class *ct_flw_http_mthd* had the highest skewness of 23.45, square root transformation was applied and the skewness was reduced to 4.00. Square root transformation was also applied to the feature *sload* and the skewness was reduced from 8.70 to 1.93.

4.4.4 Model Results

The preprocessed data is directly fed to each model; any changes in hyperparameters are discussed in the model results section of the specific model. The data is segregated based on attack type during the two-level semi-supervised label-propagation model training.

Table 4.3 Skewness of the selected features

Feature	Skewness
Proto	−3.58
Rate	3.32
Sttl	−0.68
Sload	8.70
Dload	4.69
Smean	3.73
Dmean	2.83
Ct_flw_http_mthd	23.45
Is_sm_ips_ports	7.78

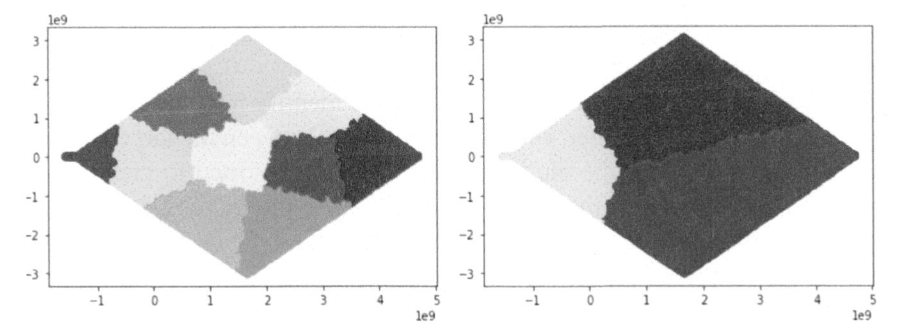

Figure 4.1 Visualization of clustering

4.4.4.1 K-Means clustering

To validate K-means clustering, the individual clusters must be analyzed for which cluster with original label can be grouped; a good indication of this is often the size of the cluster. The clusters have been represented visually via a graph in Figure 4.1

Since it is known that there are 10 different categories in the target class, the original value for K was set to 10. It is inferred from Figure 4.1 that the 10 clusters had approximately the same average size. While some of the clusters were visually smaller than others, it is visually clear that none of the classes fulfilled the 31.9% weightage that should have been seen for normal packets. Similarly, none of the classes fulfilled the 0.07% weightage for the label "Worms."

Then, an approach is executed that would work with zero-day attacks. The number of clusters is set to 2, expecting that all non-attack packets would constitute a nearly 30% cluster, while all attack packets would account for the remaining 70%. However, this split was also nearly even, similar to $K = 10$.

Finally, $K = 3$ is initialized to analyze the split for non-attack classes. This split resulted in a cluster that constituted 30% of the entire dataset. It cannot be identified which attacks were split into the other two clusters. The attacks with a smaller population in the training data seem to be split amongst both the clusters with no obvious distinguishing factor.

Thus, while a K-means clustering algorithm with $K = 3$ can identify the "Normal" data in encrypted traffic analysis, it fails to classify the malicious packets into attack types for this specific set of features.

This inability can be expected as the K-means clustering approach trusts the spatial closeness of the information in similar classes. This means that some, or several, features should be similar amongst the various data in a single cluster. However, as the correlation analysis revealed, the highest correlation magnitude for any one feature with the target was 0.3. This makes it difficult for distance-based clustering methods to identify how the clusters were formed accurately. Further, since the correlation between the target and the other features are low, the clustering method often resorted to creating similar-sized clusters during the training process. Since the training data has a high-class imbalance, the accuracy also drops.

4.4.4.2 Metrics

The data was split using stratified shuffle split to compare the supervised and semi-supervised models to ensure that all classes were evenly represented. The training data constitute 70%, and the test data comprise 30%. The original dataset is labeled. The various classification metrics analyzed are: precision, recall, F1, and accuracy.

4.4.4.3 AdaBoost

The accuracy obtained by the AdaBoost classifier is 73%. The confusion matrix in Table 4.4 shows that the classifier barely learned the target classes that had a very low number of rowsier. Impressively, however, the "Normal" class was learned perfectly.

4.4.4.4 Random forest

The accuracy obtained by the random forest classifier is 76%. Table 4.5 shows that the confusion matrix shows that the target classes with a very low number of rows were not learned well by the classifier and dropped the model's accuracy.

Table 4.4 Confusion matrix for AdaBoost

Label	Precision	Recall	F1-score	Support
Analysis	0.05	0.46	0.10	393
Backdoor	0.00	0.00	0.00	360
DoS	0.24	0.04	0.07	2370
Exploits	0.67	0.56	0.61	6772
Fuzzers	0.46	0.62	0.53	3570
Generic	0.92	0.88	0.90	8097
Normal	1.00	1.00	1.00	11,169
Reconnaissance	0.49	0.44	0.46	2098
Shellcode	0.00	0.00	0.00	233
Worms	0.00	0.00	0.00	25
Macro avg	0.38	0.40	0.37	35,069
Weighted avg	0.75	0.73	0.73	35,069

Table 4.5 Confusion matrix of random forest

Label	Precision	Recall	F1-score	Support
Analysis	0.43	0.22	0.29	417
Backdoor	0.08	0.14	0.10	350
DoS	0.26	0.58	0.36	2407
Exploits	0.66	0.41	0.51	6737
Fuzzers	0.78	0.75	0.76	3567
Generic	0.87	0.98	0.92	8023
Normal	0.98	0.94	0.96	111,246
Reconnaissance	0.94	0.65	0.77	2075
Shellcode	0.74	0.06	0.12	223
Worms	0.00	0.00	0.00	24
Macro avg	0.57	0.47	0.48	35,069
Weighted avg	0.80	0.76	0.77	35,069

As expected, the random forest ensemble classifier outperformed AdaBoost. However, the accuracy was not much better, and the less represented classes still suffered very low performance.

4.4.4.5 Semi-Supervised label propagation

The accuracy of the models is reduced significantly while deploying supervised learning models. From the confusion matrices of random forest and AdaBoost, it is clear that the most affected classes are "Analysis," "Backdoor," "Shellcode," and "Worms." These four classes achieved the worst results for both the supervised classifiers. This is due to the class imbalance.

Table 4.6 Confusion matrix for first layer model for label propagation

Label	Precision	Recall	F1-score	Support
Other	0.30	0.06	0.10	1503
DoS	0.35	0.19	0.24	3679
Exploits	0.60	0.79	0.68	10,018
Fuzzers	0.52	0.69	0.60	5455
Generic	0.93	0.98	0.96	12,000
Normal	0.94	0.93	0.94	16,800
Reconnaissance	0.71	0.13	0.22	3148
Macro avg	0.62	0.54	0.53	52,603
Weighted avg	0.76	0.77	0.74	52,603

Table 4.7 Confusion matrix for second layer model for label propagation

Label	Precision	Recall	F1-score	Support
Analysis	0.56	0.65	0.60	600
Backdoor	0.49	0.42	0.45	524
Shellcode	0.85	0.82	0.84	340
Worms	0.89	0.79	0.84	39
Macro avg	0.70	0.67	0.68	1503
Weighted avg	0.61	0.61	0.61	1503

Semi-supervised classifiers have a significant advantage over supervised classifiers and unsupervised classifiers. However, with unsupervised learning, it is difficult to guide the training. Since real-world data is not pre-labeled, the labeling task for supervised learning itself is very time-consuming. This makes retraining of the model on new data very difficult. Semi-supervised learning uses some labels to start the training process and only requires partially labeled data. Thus, it presents a good alternative for real-world applications.

The worst four classes were re-categorized into one class called "Other" to enhance training accuracy. All the data identified for the "Other" class were then fed into another classifier. Table 4.6 shows the confusion matrix for the first classification and the second model is represented in Table 4.7

The respective accuracies for the first and second models were 77% and 61%, respectively. However, the improvement in learning, especially for the second model, was incredibly significant.

In the AdaBoost classifier, "Backdoor," "Shellcode," and "Worms" had precision, recall, and F1-score values of 0.00. "Analysis" class had small magnitude of scores. In comparison, the random forest had at least some

Table 4.8 Comparing the supervised learning techniques and the proposed two-layer label propagation model

Algorithm	Accuracy	Recall
Random forest	76%	57%
AdaBoost	73%	38%
Label propagation (proposed)	77%	62%

magnitude of scores for "Backdoor" and "Shellcode." However, even random forest failed to achieve the confusion matrix scores for "Worms" above 0. Both models did assign some testing packets to "Worms," but neither could classify even one accurately.

On the other hand, the magnitude of confusion matrix scores more excellent for "Analysis," "Backdoor," and "Shellcode," and the classification of "Worms" has an 89% precision, 79% recall, and an F1-score of 0.84. This improvement is significant and proves the merit of this method. An analysis of the accuracies for the proposed model and the supervised learning models is given in Table 4.8

4.5 Discussion and Future Work

Generally, the data in encrypted network traffic analysis datasets is highly imbalanced, resulting in the development of incredibly complex systems and costly neural networks to obtain higher accuracy. However, this is not necessarily required, particularly if systems of machine learning models can achieve similar results.

The data used for the study was highly imbalanced, mostly in favor of non-attack packets. The data contained very few packets that contained "Worms" or "Shellcode." Unfortunately, this caused traditional machine learning techniques to result in lower accuracy. There was also no valid classification of the less frequent attacks. Further, unlike many popular datasets, the target data was not heavily imbalanced by 40% or more toward non-attack data. Thus, the unsupervised model could not act as a zero-day classifier or as an outlier analysis model. Feature analysis was used to select a few highly valuable features and pre-processing was utilized to remove heavy skewness in some features. This also reduced overhead and improved classification results.

The limitations of the discussed models revealed that the major issues arose due to the type of class imbalance in the data. Therefore, dealing with some magnitude of the class imbalance is expected to improve the results of the classifiers. To this end, the semi-supervised classifier is implemented in two layers, with the least frequent classes combined into one class called "Other" for the first layer. This resulted in improved confusion matrix scores for each less frequent attack. However, the most significant improvement was actually in the least frequent attack, "Worms," which the random forest and AdaBoost classifier could not classify correctly at all. Thus, the proposed semi-supervised method labels the less frequent attacks also with higher accuracy.

Multiple avenues for future research have arisen in this work. Thus, the classification results are improved by splitting the classification phase into two separate models, which have proven highly effective. In addition, the two-layer approach is analyzed to improve the accuracy of K-means to bypass the necessity for manual labeling. Further, research is required into using two-layer models for zero-day attack analysis. Finally, the plan is to test the findings of this paper on other encrypted analysis datasets to confirm that the findings are reproducible amongst varied datasets.

4.6 Conclusion

In conclusion, a thorough understanding of the data features is required to create accurate models. While supervised models are commonly used, they need the additional effort of manually labeling all of the training data. Similarly, unsupervised models are also unguided models, which may not accurately represent imbalanced data. To this end, semi-supervised models can be used to create a balance between the ease of training of unsupervised models and the control over training of supervised models. Further, a thorough analysis of the data and the effects on modeling in a supervised context allowed for an understanding of why desired results were not achieved; in the proposed model, splitting the label propagation model into two layers allowed for higher precision and recall of less frequently occurring attacks such as Worms, Shellcodes, and Backdoors.

Acknowledgment

This study was undertaken with the support of the Scheme for Promotion of Academic and Research Collaboration (SPARC) grant SPARC/2018-2019/P616/SL "Intelligent Anomaly Detection System for Encrypted Network Traffic."

References

[1] Feghhi, S., & Leith, D.J. (2016). A web traffic analysis attack using only timing information. IEEE Transactions on Information Forensics and Security, 11(8), 1747–1759.

[2] Shen, M., Liu, Y., Zhu, L., Xu, K., Du, X., & Guizani, N. (2020). Optimizing feature selection for efficient encrypted traffic classification: A systematic approach. IEEE Network, 34(4), 20–27.

[3] Barut, O., Grohotolski, M., DiLeo, C., Luo, Y., Li, P., & Zhang, T. (2020, December). Machine learning based malware detection on encrypted traffic: A comprehensive performance study. In Proceedings of the 7th International Conference on Networking, Systems and Security (pp. 45–55).

[4] Cai, X., Nithyanand, R., Wang, T., Johnson, R., & Goldberg, I. (2014, November). A systematic approach to developing and evaluating website fingerprinting defenses. In Proceedings of the 2014 ACM SIGSAC Conference on Computer and Communications Security (pp. 227–238).

[5] Cherubin, G., Hayes, J., & Juárez, M. (2017). Website fingerprinting defenses at the application layer. Proceedings on Privacy Enhancing Technologies, 2017(2), 186–203.

[6] Leroux, S., Bohez, S., Maenhaut, P. J., Meheus, N., Simoens, P., & Dhoedt, B. (2018, April). Fingerprinting encrypted network traffic types using machine learning. In Proceedings of the NOMS 2018-2018 IEEE/IFIP Network Operations and Management Symposium (pp. 1–5). IEEE.

[7] Limthong, K., & Tawsook, T. (2012, April). Network traffic anomaly detection using machine learning approaches. In Proceedings of the 2012 IEEE Network Operations and Management Symposium (pp. 542–545). IEEE.

[8] Moustafa, N., & Slay, J. (2015, November). UNSW-NB15: A comprehensive data set for network intrusion detection systems (UNSW-NB15 network data set). In Proceedings of the 2015 Military Communications and Information Systems Conference (MilCIS) (pp. 1–6). IEEE.

[9] Moustafa, N., & Slay, J. (2016). The evaluation of network anomaly detection systems: Statistical analysis of the UNSW-NB15 data set and the comparison with the KDD99 data set. Information Security Journal: A Global Perspective, 25(1-3), 18–31.

[10] Akbari, I., Salahuddin, M.A., Ven, L., Limam, N., Boutaba, R., Mathieu, B., Moteau, S., & Tuffin, S. (2021). A look behind the curtain: Traffic

classification in an increasingly encrypted web. Proceedings of the ACM on Measurement and Analysis of Computing Systems, 5(1), 1–26.

[11] Perera, P., Tian, Y.C., Fidge, C., & Kelly, W. (2017, November). A comparison of supervised machine learning algorithms for classification of communications network traffic. In Proceedings of the International Conference on Neural Information Processing (pp. 445–454). Springer, Cham.

[12] Velan, P., Čermák, M., Čeleda, P., & Drašar, M. (2015). A survey of methods for encrypted traffic classification and analysis. International Journal of Network Management, 25(5), 355–374.

[13] Cherukuri, A.K., Ikram, S.T., Li, G., Liu, X., Das, V., and Raj, A. (2021). Integrity of IoT network flow records in encrypted traffic analytics. Security and Privacy in the Internet of Things: Architectures, Techniques, and Applications, 177–205.

[14] Jha, H., Patel, I., Li, G., Cherukuri, A.K., and Ikram, S.T. (2021). Detection of tunneling in DNS over HTTPS. In Proceedings of the 2021 7th International Conference on Signal Processing and Communication (ICSC) (pp. 42–47). IEEE.

[15] Ikram, S.T., Cherukuri, A.K., Poorva, B., Ushasree, P.S., Zhang, Y., Liu, X., and Li, G. (2021). Anomaly detection using XGBoost ensemble of deep neural network models. Cybernetics and Information Technologies, 21(3), 175–188.

5

Comparative Analysis of Android Application Dissection and Analysis Tools for Identifying Malware Attributes

Swapna Augustine Nikale and Seema Purohit

Department of Computer Science, University of Mumbai, India
E-mail: swapna.nikale@gmail.com; supurohit@gmail.com

Abstract

The Android operating system is one of the most popular mobile phone operating systems. Due to its popularity and open-source nature, Android is attracted to many malware developers. They use Android mobile applications as a medium to compromise the security of the user and device. Malware developers construct malicious applications responsible for information leakage, privilege escalation, and data theft. To identify and automate the task of Android malware detection, the malware attributes should be extracted from the Android applications. This paper describes and demonstrates the tools used for Android application dissection and an analysis of malware attributes in various Android application components. The malware information that could be obtained from the Android application package is also discussed. Among the several Android application dissections and malware attribute analysis tools, it is noted that Androguard was one of the most prominent tools due to its several advantages.

Keywords: Android mobile applications, Android malware detection, Android application dissection, Android application package, Androguard, Android malware analysis tools.

5.1 Introduction

Android is one of the leading mobile operating systems, which is acceptable universally by many mobile phone users and device developers. Currently, Android is used in mobile devices and used in various appliances such as smart television, tablets, smartwatches, smart glasses, etc. The source code of the Android operating system is available to Android Open Source Projects (AOSP), an open-source initiative by Google [1]. Android's kernel is built on top of the Linux kernel's long-term support branches [2].

Android operating system, developed by Google, was used by 1.6 billion users worldwide in 2019, a 200% increase since 2012 [3]. Android captures and maintains its position as the leading mobile operating system worldwide as of June 2021 with a market share of up to 73% [4]. The official application store for Android, Google Play Store, hosts 2.89 million applications (apps) as of July 2021, after surpassing its 1 million apps in July 2013 [5].

The popularity and huge acceptance of Android mobile devices among users have attracted criminals who disrupt through various malware, exploits, or attacks. As of March 2020, the total number of new malware samples of Android amounted to 5 lakhs approximately per month [6]. These statistics present the importance of Android malware identification in Android mobile applications, which ensures security and trust in the Android community.

Malware is a piece of software that performs an unethical task on behalf of an attacker on someone's trusted system either by obtaining sensitive information and leaking it beyond the trusted environment or disrupting the services of the trusted systems. Earlier, the first malware of Android, DroidSMS, was emerging in August 2010, which had a limited scope of damage to certain application components or when a specific event was triggered. The current versions of Android malware are very stealthy and intelligent, which can bypass the detection mechanism to a certain extent and changes its behavior to something new whenever its old signature behavior gets identified by signature-based detection systems. It can also replicate itself to different variants and identify the presence of an emulator and its execution to avoid its detection.

5.2 Related Works and Present Contributions

Previous works concentrated on demonstrating the tools or tools used for malware analysis directly without explicitly stating its working scope. In [7], the author presented a few tools such as APKTool, Dex2jar, JD-GUI, VirusTotal, and Genymotion. Pan *et al.* [8] discussed that APKTool was the most popular tool among android application dissection. Rashid *et al.* [9] addressed the usage of Androguard and CuckooDroid for Android malware analysis. In [10], the authors gave a very crisp introduction on some of the tools such as APKTool, Androguard, FlowDroid, Android Tamer, and DroidBox. Garg *et al.* [11] presented an exhaustive review on Android security assessment concerning malware taxonomy, analysis techniques, code representations, and a few highlighted specific tools. Faruki *et al.* [12] concentrated on some malware analysis tools for its deployment in an experimental environment.

The requirement of working demonstration and critical and comparative analysis on malware dissection and analysis tools was much needed to select tools properly. Every tool had its unique features that would better fit in its specific malware dissection and analysis stage. This paper presents a lucid demonstration and comparative analysis on malware dissection and analysis tools that would help Android malware security researchers select the best tools appropriately at a given specific instance and experiment environment. Further study is broken into the following sections. Section 5.3 discusses the background and basic concepts about Android ecosystem, Section 5.4 elaborates Android application malware attributes and its dissection process, Section 5.5 demonstrates Android application and dissection and analysis tools, Section 5.6 presents a comparative analysis of Android APK dissection and malware attribute analysis tools, Section 5.7 includes conclusion and future work, and Section 8 presents the references.

5.3 Background and Basic Concepts of Android Ecosystem

5.3.1 Android Operating System Architecture

The Android operating system [13] is an open-source operating system built on top of the Linux kernel. The Linux kernel had been tried and tested by many developers and users throughout decades, making it one of the most secure and durable kernel systems. Some of its key security features include the following:

- Permission-based model
- Isolation of process
- Safe and secure inter-process communication (IPC)
- Removal of potentially harmful and insecure kernel components

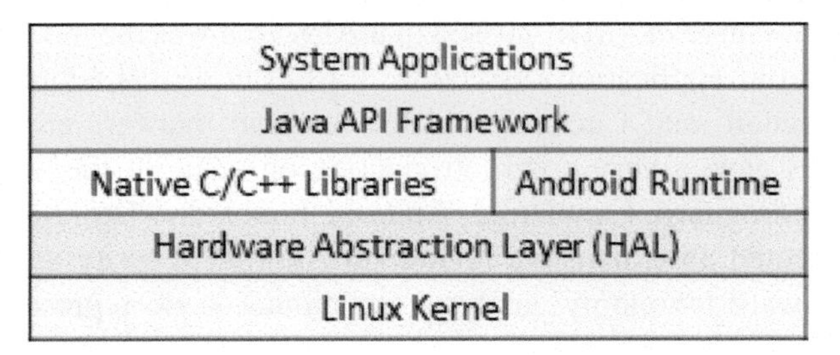

Figure 5.1 Android operating system architecture

The major components of the Android operating system include the following:

- **Linux kernel:** Linux kernel [14] is the most trusted and acceptable computing base for many security-sensitive systems. It creates a sandbox environment for applications; thus, one application does not interfere or alter the operations and contents of another application. Proper permissions should be requested by the developer of the requesting application to access information of another application. Through Security-Enhanced Linux (SELinux), Android applies access control policies and establishes mandatory access control (MAC).
- **Hardware abstraction layer (HAL):** HAL consists of multiple libraries that provide an interface to every specific type of hardware component such as a microphone, camera, Bluetooth, etc. When an application programming interface (API) framework requests to access hardware, the respective library module is loaded to assist in its usage.
- **Android runtime (ART):** Every application process has its instance on the ART. It also assists in running multiple virtual memories in resource-constrained devices by executing a bytecode format DEX file designed specifically for Android. Some of the notable features of ART are as follows [15]:
    - Ahead-of-time (AOT) compilation
    - Improved garbage collection
    - Improved app development and debugging
    - Improved diagnostic details in exceptions and crash reports
- **Native C/C++ libraries:** Some of the core components of the Android system, such as HAL and ART, require native libraries for its execution which is written in C/C++.

- **Java API framework:** Java API forms a building block for creating Android apps. It assists in simplifying coding through code reuse, modularity, etc. The complete set of API in Android OS is written in Java language.
- **System applications:** Android applications (apps) represent the user interface (UI) through which the user can perform any tasks and interact with the operating system. Android apps can be installed ones (from a trusted app store or third-party app store) or core apps (installed by default either by hardware manufacturers or Android OS developers).

5.3.2 Android Application Fundamentals [16]

Android application (app) is software designed to run on an Android device or any emulators. Android applications can be written in Java, Kotlin, or C++ languages. The Android application package (APK) is a compressed package containing the files necessary for Android applications during their runtime.

Android apps exist in their security sandbox. Thus, any untrusted or unauthorized data access among applications cannot happen. Some of the security features for Android apps are listed below [17]:

- Each app is trusted as a different user in the Android operating system.
- Every app has its own Linux process, which is created whenever the app's components are executed and terminated when the app no longer needs it.
- Due to Android's "principle of least privilege," the Android app has access only to necessary components for its execution. Any necessity of data sharing among applications makes it possible based on one of the two scenarios given below:
    - Apps signed using the same certificate, which come from the same developer, can be assigned using the same user ID and run on the same Linux process ID. This leads to resource conservation.
    - If an application does not share the same certificate (in most cases), it can request permissions to access user-sensitive information related to the device or app. Permissions are granted explicitly by the user.

Some of Android applications' components that are necessary for building Android apps are listed below:

- **Activities:** It is an entry point for interacting with users. It represents a single screen with a user interface. For example, an alarm app might have one activity for creating an alarm and another activity for deleting it.

- **Services:** It is a general-purpose entry point that assists in keeping the app running in the background for long-running operations or any remote processes; for example, keeping the music playing in the background, receiving an alarm, or other app notifications. It does not provide a user interface.
- **Broadcast receivers:** It is a component that enables the system to deliver events to apps that are not currently active. For example, the screen turns off when the device is not in use, battery is low, or there is no background notification to the user or app.
- **Content providers:** It manages the shared set of app data stored in a file system, SQLite database, web, or any other persistent storage location. With the help of the content provider, other apps can access, query, and modify the data only if the content provider grants permission to it.

5.4 Android Application Malware Attributes and its Dissection Process

5.4.1 Android Application Malware Attributes

Listed below are some of the malware attributes which were observed in the APK files [9], [12], [18].

- assets/ – It includes files like text, XML, fonts, music, and video in applications and allows to access the data which was raw or untouched. This directory stores some .dex files or any other payloads executed during dynamic loading.
- lib/ – It includes compiled code and several sub-directories, each available for a specific platform type. The lib directory shows some of the chances of customized and malicious libraries occurrence, loaded during app execution.

  Listed below are other files included in this directory:

  - MANIFEST.MF – It is a text file that contains base 64 encoded SHA-1 hashes of all files in APK.
  - CERT.SF – Base 64 encoded SHA-1 hashes of corresponding lines in MANIFEST.MF file.
  - CERT.RSA – Contains public signatures of the application developer, which is used for validating the contents.

Figure 5.2 Android application package

- res/ – This directory includes some of the non-compiled resources that the application could access using resource_ID. All resource IDs are defined in the "R" class of the project. Since the res directory stores some of the non-compiled resources, malicious payloads could be hidden and uploaded during dynamic loading.
- AndroidManifest.xml – It includes essential information about applications like entry points, package name, components, permissions, minimum level of Android API, libraries, intents, etc. Due to the numerous entry points, a backdoor could be developed, which could cause the entry of some other malicious applications. Permission escalation, blacklisted package names, activities, and services requested by the app could be obtained from this file, which helps understand its original intention. A maximum amount of static malware attributes could be obtained from this file [8].
- classes.dex – It contains .java/.class files in DEX format, which helped in the brief understanding of the actual intention of an application. It was also sometimes helpful in malware signature generation and used to find obfuscation occurrence in very few cases.

- resources.arsc – It comprises compiled resources, language strings, styles, layout files, images, and the path to content that are not linked to resources.arsc. This file could contain some malicious payloads loaded during dynamic execution. Natural language processing techniques could also be applied to language strings to extract malware-like behavior in the application.

5.4.2 Android Application Malware Dissection

The process of separating different file components from an Android application package (APK) is called Android application dissection. It is considered the first step of analyzing malware presence in Android apps [19]. The process of Android application dissection consists of the following three main processes [20].

- **Decompressing APK:** An Android application package (APK) is a compressed file format containing several files needed for the execution of an application. To analyze the Android application for any malware attributes, it has to be decompressed so that individual file components could be obtained and analyzed. Tools like Unzip, 7zip, and APKTool could be used for decompressing APK [12].
- **Decompiling .dex/.java files:** Once the APK file was decompressed, its classes.dex file is exposed. The .dex file could be decompiled further to obtain several .jar, .class, or .java files. Tools like Jadx, JD-GUI, Smali/Backsmali, Dex2Jar, Enjarify, and Dedexer could be used for decompiling .dex files, interpreting .java files and .smali files, converting .dex files to .jar files, and accessing several .class files from the .jar files [21].

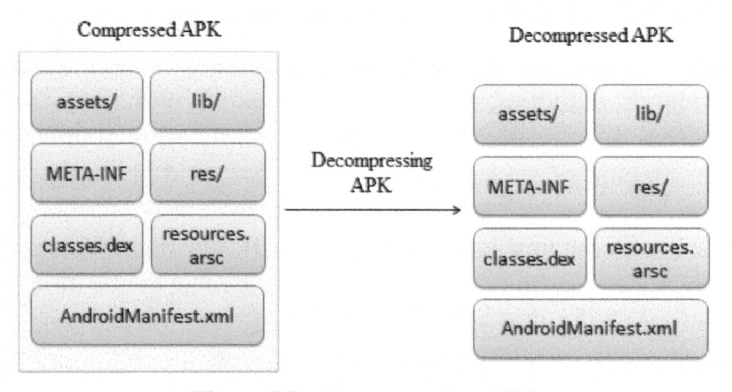

Figure 5.3 Decompressing APK

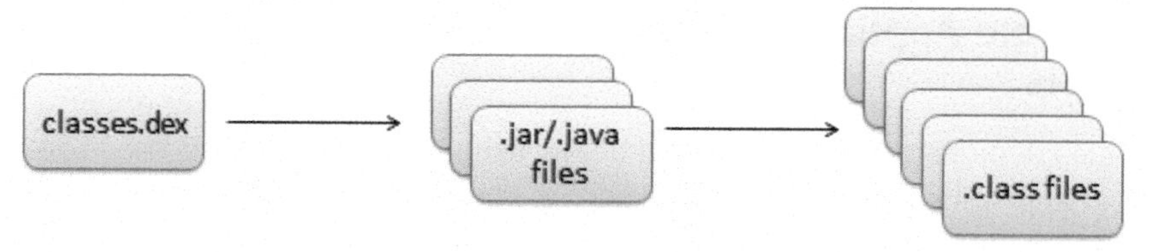

Figure 5.4 Decompiling .dex/.java files

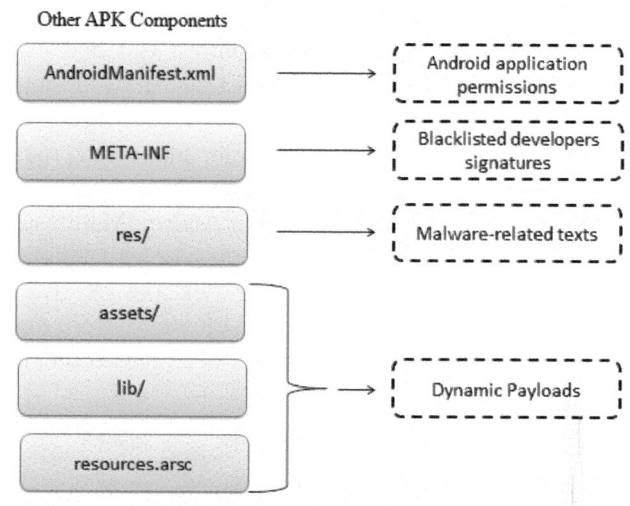

Figure 5.5 Other APK components

- **Analyzing other APK components:** Other APK components such as AndroidManifest.xml, META-INF, lib/, resources.arsc, res/, and assets/ could also be used for analyzing the presence of any malware tracks. The most popularly used component for Android malware analysis is AndroidManifest.xml. Since Android is built on the permission-based model, AndroidManifest.xml files include permissions necessary for the app execution. This attribute contributed to large information concerning malware behaviors. Strings from the res directory could be analyzed through natural language processing (NLP) to obtain malware-related texts such as LOTTERY, VIRUS, ALERT, etc. The developer signature presented in META-INF could be used to check if it belonged to blacklisted developers of previous malware spread history or sometimes to identify the presence of malware families or any other malware variants. Other components such as assets/, lib/, and resources.arsc would contain some traces of dynamic payloads.

5.5 Android Application Dissection and Malware Analysis Tools

A detailed demonstration is presented below using some of the tools used for Android application dissection and Android malware attribute identification using one malware sample from the DREBIN dataset [22].

- **Unzip/7zip:** This basic decompressing tool is used to unzip Android APK files. After decompressing the APK, the Android components such as assets/, META-INF, res/, AndroidManifest.xml, classes.dex, and resources.arsc were obtained. It was noted that AndroidManifest.xml was not obtained in a readable format.
- **APKTool:** APKTool [23], a popularly used decompressing tool, is used for decompressing Android APK and provides all Android components in a much readable format. It was noted that AndroidManifest.xml was in a readable format and presented Dalvik bytecode in .smali format.
- **Jadx:** This tool [24] assists in decompiling .dex files to .java. It could produce source code from either Android APK or Android classes.dex files. It could also decode AndroidManifest.xml and also other resources

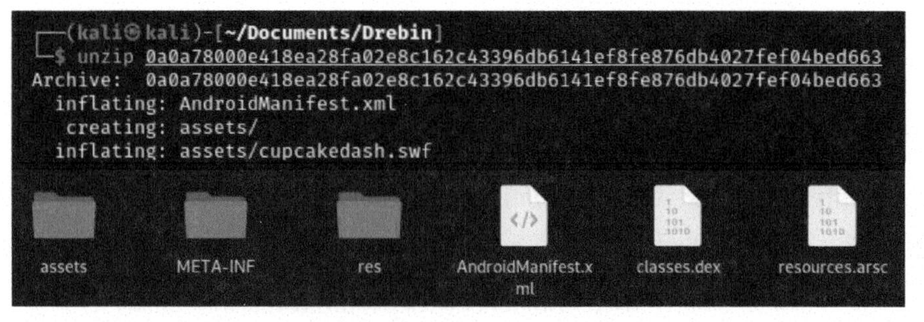

Figure 5.6 Unzip

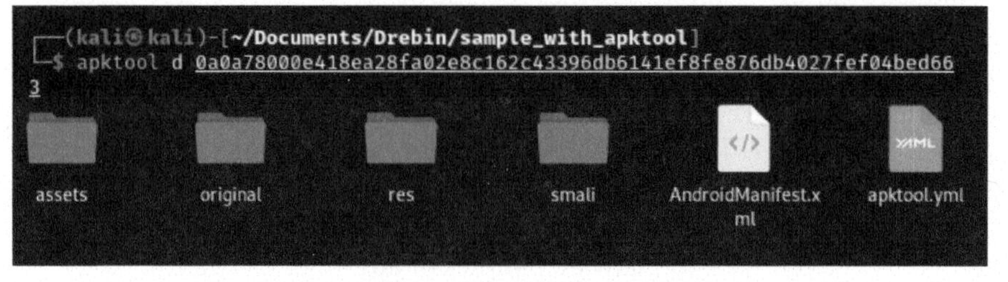

Figure 5.7 APKTool

```
┌──(kali㉿kali)-[~/Documents/jadx-1.2.0/bin]
└─$ jadx -d /home/kali/Documents/Drebin/New/ /home/kali/Documents/Drebin/New/PocketMine.apk
```

```
┌──(kali㉿kali)-[~/Documents/jadx-1.2.0/bin]
└─$ jadx -d /home/kali/Documents/Drebin/New/ /home/kali/Documents/Drebin/New/0a0a
78000e418ea28fa02e8c162c43396db6141ef8fe876db4027fef04bed663
```

Figure 5.8 Jadx

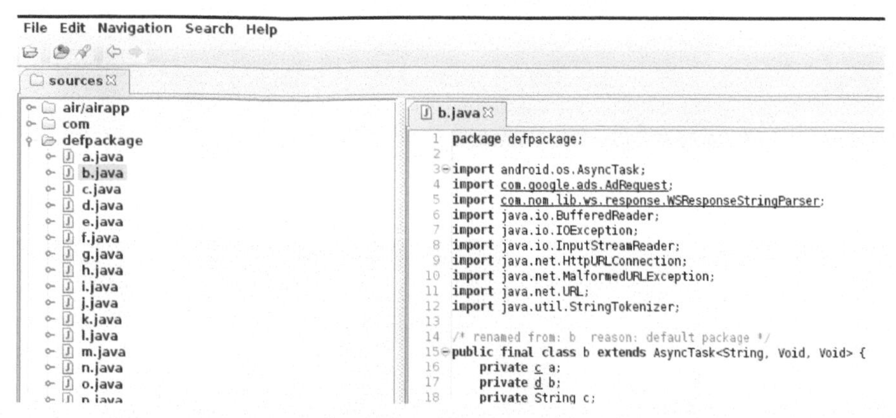

Figure 5.9 JD-GUI

from resources.arsc. In Figure 5.8, Jadx was demonstrated using PocketMine APK [25].

- **Java Decompiler – Graphical User Interface (JD-GUI):** This tool [26] can only interpret .java files. Thus, it should be used in combination with other DEX decompiling tools. It was noted that when an APK file was decompressed using APKTool, .smali files were generated, which cannot be analyzed using JD-GUI. The classes.dex file obtained after decompressing an APK using Unzip was also not in a supported format for analyzing through JD-GUI. To make it accessible, the classes.dex file was converted to several .java files using the dedexing process.
- **Smali/Baksmali:** Smali and Baksmali [27] is a combo tool that performs assembler and disassembler tasks, respectively. Smali can assemble several .smali files to classes.dex format, whereas Baksmali can disassemble an application's classes.dex to multiple fragments of .smali files. The Java source code in .smali format was in a readable format.
- **Dex2Jar:** This tool [28] converts classes.dex files to .jar files. The .jar files could be extracted to obtain a .class file, a readable file.

```
┌──(kali㉿kali)-[~/Documents/Drebin/baksmali]
└─$ baksmali disassemble /home/kali/Documents/Drebin/baksmali/PocketMine.apk -o /home/kali/Documents/Drebin/baksmali/PocketMine
Picked up _JAVA_OPTIONS: -Dawt.useSystemAAFontSettings=on -Dswing.aatext=true

┌──(kali㉿kali)-[~/Documents/Drebin/baksmali]
└─$ smali assemble /home/kali/Documents/Drebin/baksmali/PocketMine -o /home/kali/Documents/Drebin/baksmali/classes.dex
Picked up _JAVA_OPTIONS: -Dawt.useSystemAAFontSettings=on -Dswing.aatext=true
```

Figure 5.10 Smali/Baksmali

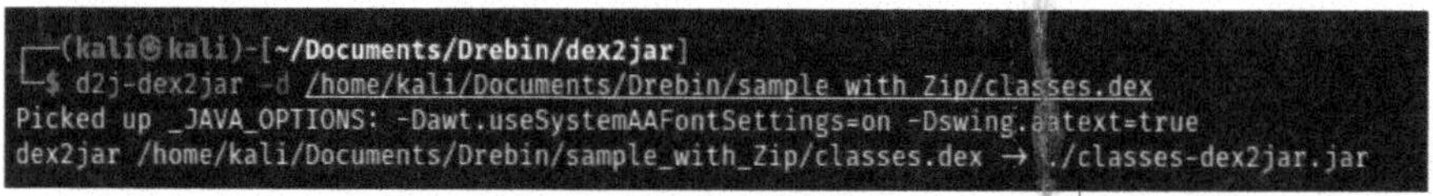

```
┌──(kali㉿kali)-[~/Documents/Drebin/dex2jar]
└─$ d2j-dex2jar -d /home/kali/Documents/Drebin/sample_with_Zip/classes.dex
Picked up _JAVA_OPTIONS: -Dawt.useSystemAAFontSettings=on -Dswing.aatext=true
dex2jar /home/kali/Documents/Drebin/sample_with_Zip/classes.dex -> ./classes-dex2jar.jar
```

Figure 5.11 Dex2Jar

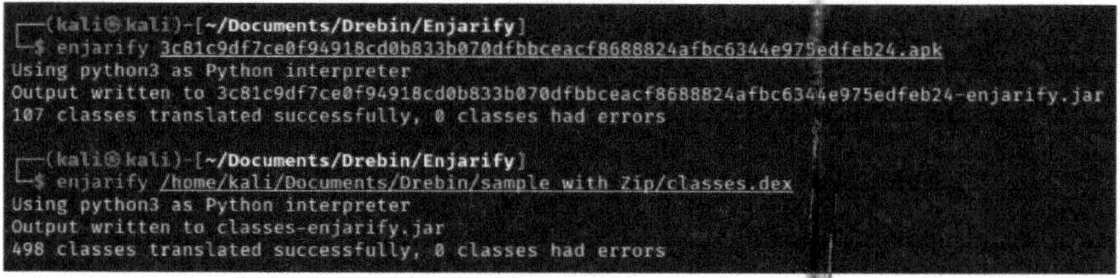

```
┌──(kali㉿kali)-[~/Documents/Drebin/Enjarify]
└─$ enjarify 3c81c9df7ce0f94918cd0b833b070dfbbceacf8688824afbc6344e975edfeb24.apk
Using python3 as Python interpreter
Output written to 3c81c9df7ce0f94918cd0b833b070dfbbceacf8688824afbc6344e975edfeb24-enjarify.jar
107 classes translated successfully, 0 classes had errors

┌──(kali㉿kali)-[~/Documents/Drebin/Enjarify]
└─$ enjarify /home/kali/Documents/Drebin/sample_with_Zip/classes.dex
Using python3 as Python interpreter
Output written to classes-enjarify.jar
498 classes translated successfully, 0 classes had errors
```

Figure 5.12 Enjarify

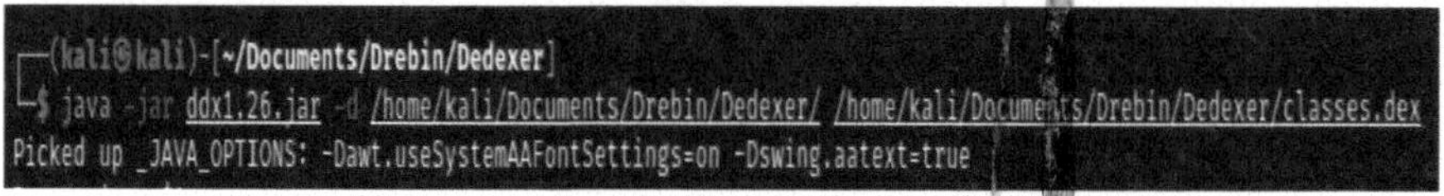

```
┌──(kali㉿kali)-[~/Documents/Drebin/Dedexer]
└─$ java -jar ddx1.26.jar -d /home/kali/Documents/Drebin/Dedexer/ /home/kali/Documents/Drebin/Dedexer/classes.dex
Picked up _JAVA_OPTIONS: -Dawt.useSystemAAFontSettings=on -Dswing.aatext=true
```

Figure 5.13 Dedexer

- **Enjarify:** This tool [29] helps in converting the Dalvik bytecode (.dex) or Android application package (.apk) to Java bytecode (.jar). By accessing the .jar files, all .class files associated with an APK were accessible.
- **Dedexer:** Dedexer [30] is a disassembler tool that can read .dex files and turn them into an "assembly like format," largely influenced by Jasmin syntax but contained Dalvik opcodes. It was noted that the contents of classes.dex file was in a readable format.
- **BytecodeViewer:** This [31] tool is specifically used for viewing source codes from an APK directly or from other files of formats, such as .jar, .zip, .class, .dex, etc. It also helps in viewing the source codes

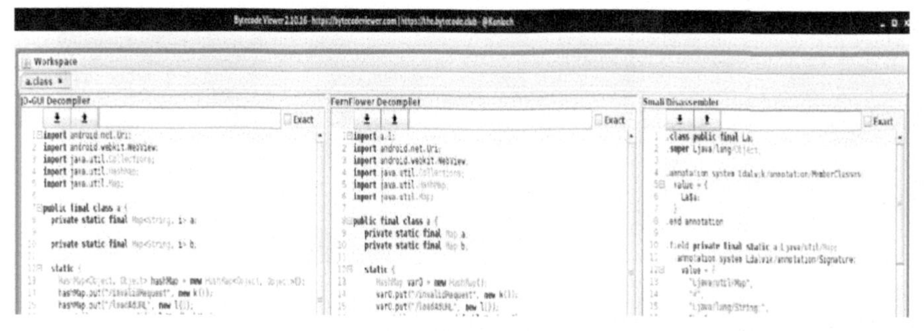

Figure 5.14 BytecodeViewer

Figure 5.15 Android SDK

using three different compilers simultaneously on a single screen. This helps in comparing source code in a decompiled format in a minimum of three different decompilers. This tool also allowed navigation to different parts of the source code based on search terms using the class name, method name, string, function name, etc. In Figure 5.14, JD-GUI, Fernflower, and Smali decompilers' output was presented.

- **Android SDK:** It [32] is a software development kit released by Google specifically for Android. It was designed with the intention of APK development and for analyzing APK files. All the components of an APK could be analyzed using this tool.
- **Androguard:** It [33] is a python-based tool used to analyze Android files. Among all the tools discussed and demonstrated, Androguard is considered the most feasible and user-friendly tool [9]. The only restriction it posed was working only with Python 3. It acts as a disassembler and decompiler tool that could work with APK or classes.dex files. It includes commands for analyzing APK for information such as permissions, activities, package name, app name, the path of an app icon, numeric version, version string, minimal, maximal, target, and SDK versions, decoded XML, and also access every class of an application.

```
┌──(kali㉿kali)-[~/Documents/Drebin/JD-GUI (DO NOT OPEN THE APK)]
└─$ androguard analyze
Androguard version 3.4.0a1 started
In [1]: a, d, dx = AnalyzeAPK("3c81c9df7ce0f94918cd0b833b070dfbbceacf8688824afbc6344e975edfeb24.apk")

NameError                                 Traceback (most recent call last)
/usr/lib/python3/dist-packages/androguard/cli/main.py in <module>
----> 1 a, d, dx = AnalyzeAPK("3c81c9df7ce0f94918cd0b833b070dfbbceacf8688824afbc6344e975edfeb24.apk")

NameError: name 'AnalyzeAPK' is not defined

In [2]: from androguard.misc import AnalyzeAPK

In [3]: a, d, dx = AnalyzeAPK("3c81c9df7ce0f94918cd0b833b070dfbbceacf8688824afbc6344e975edfeb24.apk")
```

```
In [5]: a.get_permissions()

In [6]: dx.get_classes()
```

Figure 5.16 Androguard

Table 5.1 Comparative analysis of Android APK dissection and malware attribute analysis tools

Tools	Actions			Access to Android APK files						
	Decompressing APK	Decompiling DEX	Analyzing APK components	AndroidManifest.xml	assets/	lib/	META-INF	res/	resources.arsc	classes.dex
Unzip/7zip	✓		✓	✓	✓	✓	✓	✓	✓	✓
APKTool	✓		✓	✓	✓	✓		✓		✓
Jadx	✓	✓	✓(only .dex)	✓						✓
JD-GUI			✓(only .java)							✓
Smali/Baksmali		✓	✓(only .dex)							✓
Dex2Jar		✓	✓(only .dex)							✓
Enjarify		✓	✓(only .dex)							✓
Dedexer		✓	✓(only .dex)							✓
BytecodeViewer	✓	✓	✓(only .java)							✓
Android SDK	✓	✓	✓	✓	✓	✓	✓	✓	✓	✓
Androguard	✓	✓	✓	✓	✓	✓	✓	✓	✓	✓

Another advantage of the tool is that the current work could be saved using sessions whenever a large APK was analyzed and returned later. Jadx was also embedded with Androguard, which could help decompile .dex files. Bulk analysis of APK was also possible through Androguard.

A comparative study on the discussed tools was undergone and presented in Table 5.1, which would help Android malware reverse engineers select the best tool based on the task to be performed and files available at the specific instance. It was noted that Androguard, a python-based tool, could assist in automatic malware attributes extraction using python scripts and work with all the files of an Android APK. Thus, it could be a preferred tool in comparison with Android SDK.

5.6 Conclusion and Future Work

A comparative study would help Android malware reverse engineers select the best tool based on the task to be performed and files available at the specific instance. It was noted that Androguard, a python-based tool, could assist in automatic malware attributes extraction using python scripts and work with all the files of an Android APK. Thus, it could be a much preferred tool in comparison with Android SDK. Malware authors target Android systems due to their open-source nature and large customer usage. To understand and automate malware detection in Android applications, extracting and analyzing the various malware attributes is mandatory. Several tools are available in the market, which serve this purpose. It is noted that Androguard collects Android malware attributes precisely and analyzes most of the Android components, which helps malware researchers to collect a wide range of malware attributes. Since every tool performs the tasks in different ways and creates an output file of different formats, a single universal platform for such tools in the future would be much benefitted.

References

[1] "Android Open Source Project." https://source.android.com/
[2] "Android Common Kernels." https://source.android.com/devices/architecture/kernel/android-common
[3] "Population of internet users worldwide from 2012 to 2019, by operating system." https://www.statista.com/statistics/543185/worldwide-internet-connected-operating-system-population/
[4] "Mobile operating systems' market share worldwide from January 2012 to June 2021." https://www.statista.com/statistics/272698/global-market-share-held-by-mobile-operating-systems-since-2009/
[5] "Number of available applications in the Google Play Store from December 2009 to July 2021." https://www.statista.com/statistics/266210/number-of-available-applications-in-the-google-play-store/
[6] "Development of new Android malware worldwide from June 2016 to March 2020." https://www.statista.com/statistics/680705/global-android-malware-volume
[7] D. Mugisha, "Android application malware analysis," p. 29.
[8] Y. Pan, X. Ge, C. Fang, and Y. Fan, "A systematic literature review of Android malware detection using static analysis," IEEE Access, vol. 8, pp. 116363–116379, 2020, doi: 10.1109/ACCESS.2020.3002842.
[9] W. Rashid, D. N. Gruschka, F. Kiel, L. Bernhard, and Q. GmbH, "Automatic Android malware analysis," p. 58.
[10] C. Negi, P. Mishra, P. Chaudhary, and H. Vardhan, "A review and case study on Android malware: Threat model, attacks, techniques

and tools," Journal of Cyber Security and Mobility, Mar. 2021, doi: 10.13052/jcsm2245-1439.1018.

[11] S. Garg and N. Baliyan, "Android security assessment: A review, taxonomy and research gap study," Computers & Security, vol. 100, Jan. 2021, Art. no. 102087, doi: 10.1016/j.cose.2020.102087.

[12] P. Faruki *et al.*, "Android security: A survey of issues, malware penetration, and defenses," IEEE Communications Surveys & Tutorials, vol. 17, no. 2, pp. 998–1022, 2015, doi: 10.1109/COMST.2014.2386139.

[13] "Platform architecture." https://developer.android.com/guide/platform

[14] "System and kernel security." https://source.android.com/security/overview/kernel-security#linux-security

[15] "Android runtime (ART) and Dalvik." https://source.android.com/devices/tech/dalvik#features

[16] "Application fundamentals." https://developer.android.com/guide/components/fundamentals

[17] R. Mayrhofer, J. V. Stoep, C. Brubaker, and N. Kralevich, "The Android platform security model," ArXiv190405572 Cs, Dec. 2020, Accessed: Sep. 26, 2021. [Online]. Available: http://arxiv.org/abs/1904.05572

[18] Z. Wang, Q. Liu, and Y. Chi, "Review of Android malware detection based on deep learning," IEEE Access, vol. 8, pp. 181102–181126, 2020, doi: 10.1109/ACCESS.2020.3028370.

[19] Y. Zhou and X. Jiang, "Dissecting Android malware: Characterization and evolution," in 2012 IEEE Symposium on Security and Privacy, San Francisco, CA, USA, May 2012, pp. 95–109. doi: 10.1109/SP.2012.16.

[20] W. Wang *et al.*, "Constructing features for detecting android malicious applications: Issues, taxonomy and directions," IEEE Access, vol. 7, pp. 67602–67631, 2019, doi: 10.1109/ACCESS.2019.2918139.

[21] S. Hutchinson and C. Varol, "A survey of privilege escalation detection in Android," in 2018 9th IEEE Annual Ubiquitous Computing, Electronics & Mobile Communication Conference (UEMCON), New York City, NY, USA, Nov. 2018, pp. 726–731. doi: 10.1109/UEMCON.2018.8796550.

[22] D. Arp, M. Spreitzenbarth, M. Hübner, H. Gascon, and K. Rieck, "Drebin: Effective and explainable detection of Android malware in your pocket," presented at the Network and Distributed System Security Symposium, San Diego, CA, USA, 2014. doi: 10.14722/ndss.2014.23247.

[23] "APKTool." https://ibotpeaches.github.io/Apktool/

[24] "Jadx." https://github.com/skylot/jadx

[25] "PocketMine." https://m.apkpure.com/pocket-mine-2/ca.roofdog.pm2
[26] "JD-GUI." https://github.com/java-decompiler/jd-gui
[27] "Smali/Baksmali." https://github.com/JesusFreke/smali
[28] "Dex2Jar." https://github.com/pxb1988/dex2jar
[29] "Enjarify." https://github.com/google/enjarify
[30] "Dedexer." https://sourceforge.net/projects/dedexer/
[31] "BytecodeViewer." https://github.com/Konloch/bytecode-viewer
[32] "Android SDK." https://developer.android.com/studio
[33] "AndroGuard." https://github.com/androguard/androguard

6

Classifying Android PendingIntent Security using Machine Learning Algorithms

Pradeep Kumar D. S. and Geetha S.

School of Computer Science and Engineering, VIT Chennai, India
E-mail: pradeepkumarst@gmail.com; geetha.s@vit.ac.in

Abstract

The Android app uses PendingIntent to delegate authority to other potential apps temporarily. Though this temporary delegation adds flexibility to the software design by collaborating with any third-party apps dynamically, it also becomes a caveat when malware acquires the PendingIntent. Unprotected public broadcasts and PendingIntents created with an empty base intent are some of the vulnerable features that malware utilizes to perform unauthorized access and privilege escalation attacks on the delegated PendingIntent. By leveraging the machine learning techniques for security predictions, our research could assist cyber forensic tools in identifying the dynamic vulnerabilities present in an application. This paper presents four machine-learning-aided approaches for static analysis of the mobile applications: (1) based on the PendingIntent Flag usages, and (2) based on the type of broadcast across apps. Our approach provides a method for automated static code analysis and malware detection with better accuracy and helps to reduce the app security analysis time. Our app classification achieved an *F*-score of 76.6%.

Keywords: PendingIntent, intent analysis, Android, information flow control, privilege escalation, unauthorized intent receipt, random forest, logistic regression, KNN, J48.

6.1 Introduction

Applications delegate authorities that they own to other apps by sharing PendingIntent (PI) [6] by wrapping it inside an intent object (a.k.a. WrappingIntent (WI) [27]). Android apps use PI to make dynamic delegation of activities across components. It could be within the same app or across apps. By using PI, the PI creating app grants all its privileges. It includes the permission it acquired and its identity to the receiver app. The PI is temporary permission shared between components that will be automatically revoked once the receiver app performs the requested operation. Thus, the PendingIntent receiver acts on behalf of the sender by utilizing the sender's permissions and identity. In Android, sharing between components is called inter-component communication (ICC) – exchanging intents between components. Internally, the ICC can be classified as *implicit* or *explicit communication*. In implicit communication, the intent specifies a generic action instead of a definite receiving component's name. The OS's responsibility is to identify the matching receiver based on the specified action.

On the other hand, by using explicit communication, the intent specifies the receiving component identity explicitly. Thereby, components other than the specified identity cannot receive the intent. A study on 7406 applications by Feizollah *et al.* [9] found that 91% use intents to communicate across components, where 28.78% are implicit and 71.22% are explicit intents. Explicit intents are safe by confining the communication with known components, and the implicit intents (a.k.a. public broadcast) facilitate easy collaboration with third-party apps. However, it is vulnerable to malware attacks [16].

PendingIntent exchanged through implicit intents are susceptible to malware attacks, where malware (an app with less privilege compared to the sender app) can sniff the exchanged intent just by registering to the specific intent action. Thereby, malware app consumes the PendingIntent's information and the sender's permissions and identity. In this paper, we make a novel mechanism of identifying the PI attacks by inferring various properties of benign and malware apps using some famous ML algorithms like KNN, random forest, Naïve Bayes, logistic regression, and J48 decision trees.

6.2 Threat Model

Android, by default, grants apps with minimal privileges, and additional permissions like reading the GPS or making phone calls are granted upon explicit request through AndroidManifest.xml. Malicious apps usually request minimal permissions and collude with privileged apps through ICC. PendingIntents exchanged through PublicBroadcast can be sniffed by malware by registering to the corresponding broadcast actions. Thus, a failure to guard the public broadcast with a dynamic verification of the receiver's identity can lead to threats like unauthorized intent receipt (UIR) and privilege escalation (PE).

For example, an event planner app (e.g., Business Calendar) interacting with third-party location-based notification apps (e.g., Flipp) could delegate responsibility to notify the events. Thus, the above security threats happen when the system fails to verify the capability of the sender/receiver dynamically.

Public broadcasts are vulnerable to passive eavesdropping or active denial of service attacks [16]. For example, a malicious receiver can register for the same actions and categories as a benign component and sniff the public broadcast intents resulting in an UIR attack. Similarly, in PE (or confused deputy), a malicious component can exploit the vulnerable interfaces of a benign app with a broadcast carrying vulnerable PendingIntent.

The PI is vulnerable to attacks when created without best-coding practice, as shown in Figure 6.1.

The behavior of a PI is determined based on the flags used to create it. Similarly, these flags determine the possible attacks on the PI, as explained below, for example:

1. FLAG_ONE_SHOT indicates that the PendingIntent can be used only once. This kind of PendingIntents can be attacked by malware just by invoking them before the benign app does; thereby, the benign receiver cannot use the same PendingIntent again. Thus, a non-deterministic behavior is created in the workflow.

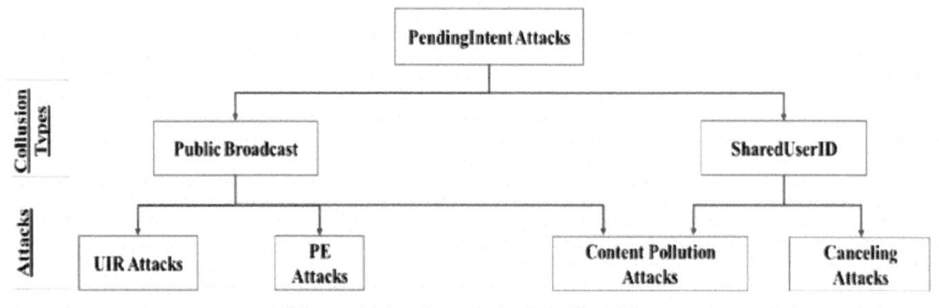

Figure 6.1 PI attack classification

2. In FLAG_UPDATE_CURRENT, the PendingIntent extras can be maliciously modified by the malware, resulting in an undesired workflow sequence (can lead to content pollution attacks).
3. FLAG_MUTABLE is used with inline reply or bubbles, where the underlying base intent is modifiable. When a malware sniffs such kinds of PendingIntents, it can modify the base intent (can lead to content pollution attacks).
4. FLAG_CANCEL_CURRENT and FLAG_NO_CREATE are the flags used only at the creation time and do not specify PendingIntents' behavior, hence having no impact on the attacks (can lead to content pollution attacks).
5. The FLAG_IMMUTABLE gives you strong protection from unwanted modifications by other receiver apps. However, when the malicious colluding app shares user ID with the PendingIntents creating/source app, it can perform canceling or retargeting PI attacks on the original PendingIntent.

   When the created PI is broadcasted through public broadcast, the following types of attacks occur:
6. If the PI is created with an empty base intent (i.e., constructed without intent action or explicit target component) or if the base intent uses implicit actions instead of an explicit receiver, it can lead to privilege escalation (PE) attack.
7. If the PI is exchanged through an unprotected broadcast channel, it can lead to an unauthorized intent receipt (UIR) Attack. Using Android's protected broadcast like Signature or SignatureOrSystem permission or mandatory receiver permissions [7], the app can avoid UIR attacks.

   Similarly, if the PI creating app and the malware shares the same user ID (using Android's SharedUserId), it can lead to the following attack:
8. The shared UID malware can cancel the PI.

6.2.1 Observations

Some existing research studies on PendingIntent security threats are RAICC [20] and PIAnalyzer [27]. RAICC instruments the PendingIntent calls (AICC calls) by adding a method startActivity() with the right intent as a parameter, thereby converting them to standard ICC calls that can be further processed with IccTA [21] and Amandroid [18]. PIAnalyzer statically analyzes PendingIntent vulnerabilities. We found that designing an application layer pluggable library with a strong theoretical foundation can help the developer understand the application-level attacks like UIR and PE. It helps to control the dynamic collaboration with other apps, which is currently unavailable.

6.2.2 Our Contributions

In this paper, we present a real-time machine-learning-based analysis model that can extract the required PendingIntent information from the APK and classify the application into predefined security classes. Our model analyzes the following two security features:

1. PI flag analysis – From the .apk file (Android application), we statically extract the flags used to create PendingIntent.
2. Broadcast analysis – The model can classify the ICC broadcast as public and protected, thereby predicting the application's security class.

Classifying security attacks based on PendingIntent can help the security analyst model the uncertain dynamic environment, where an app can hand over authority to a third-party app that may breach the integrity of the intended security. As proof of our mechanism, we have implemented Murax, which can extract the required information from the binary and feed automatically to our ML model.

6.3 Data Collection and Pre-processing

PendingIntent is one of the recent security areas and, therefore, lacks a standard dataset to train the ML models. This motivates us to build a tool to extract the PI information from the available Android APK benchmark repository. Figure 6.2 explains the process involved in dataset creation and the pre-processing steps involved in cleaning the data.

6.3.1 Dataset Discussion

This section presents our empirical study and our discovery on the PendingIntent usages in real-world apps.

For our study, we use the following dataset: (1) CICMalDroid-2020 dataset [37], a collection of 17,241 Android apps' samples spanning between five distinct categories: Adware, Banking malware, SMS malware, Riskware, and Benign, and extensive samples from several sources including VirusTotal service, Contagio security blog, AMD, MalDozer, and other datasets used by recent research contributions; (2) the Drebin dataset [38] that contains 5560 applications from 179 different malware families; and (3) 1121 Google Play

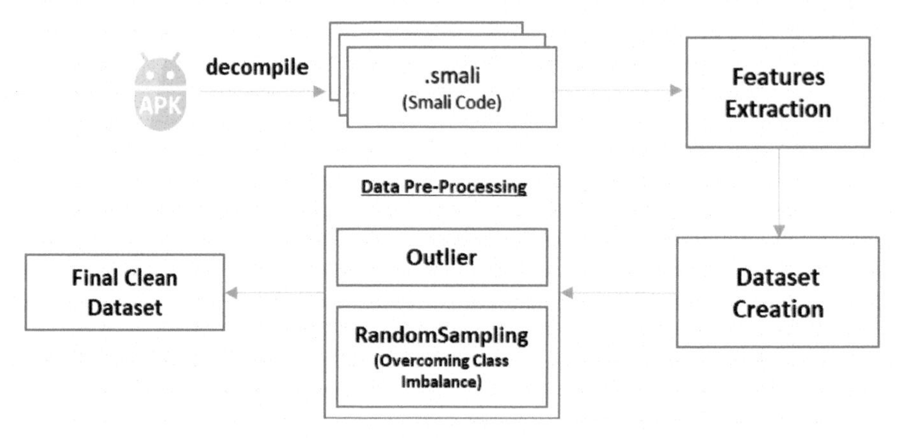

Figure 6.2 Data extraction and pre-processing

Dataset	#App	#C	#M	#I	#PI	1s	Nc	Cc	Uc	Im	PubBr	VulPI	VulWr	#ProBr	TotBr	%VulPI	%VulWr	%ProBr	%TotVulBr	%TotVulPI
Adware	1514	568168	3261490	96249	11836	43	23	2558	4158	119	1351	4805	6	55	4068	40.60	0.44	1.35	33.21	40.65
SMS	4822	153920	755568	61556	14032	2	1	6094	303	0	6022	7368	0	2	6024	51.80	0.00	0.03	99.97	51.80
Banking	2507	655536	3387150	62551	7037	810	13	578	1762	3	1583	2026	2	103	1686	28.79	0.13	6.11	93.89	28.82
Benign	4038	10161456	45900788	474921	25245	1658	500	3142	10009	58	9756	9472	163	855	10611	37.52	1.67	8.06	91.94	38.17
Riskware	4363	2239942	11482168	124183	12386	818	40	625	5113	8	2752	2248	0	408	3160	18.15	0.00	12.91	87.09	18.15
Drebin	5560	1319375	7084474	166213	15842	349	92	3394	2043	12	3437	6208	77	1	7522	39.19	2.24	0.01	45.69	39.67
Google Play	1123	6280234	25689943	94573	8442	628	423	650	2915	374	1048	2762	8	127	6311	32.72	0.78	2.01	16.61	32.81
TOTAL	**23927**	**21378631**	**97561581**	**1080246**	**94820**	**4308**	**1092**	**17041**	**26303**	**574**	**25949**	**34789**	**256**	**1551**	**39382**	**36.69**	**0.99**	**3.94**	**65.89**	**36.96**

NOTE:
1s - FLAG_ONE_SHOT
Nc - FLAG_NO_CREATE
Cc - FLAG_CANCEL_CURRENT
Uc - FLAG_UPDATE_CURRENT
Im - FLAG_IMMUTABLE
PI - PendingIntent
C - Class
PubBr - Public Broadcast
VulPI - Vulnerable PI Creation
I - Intent
VulWr - Vulnerable PI Transfer (WrappingIntent)
%VulPI - Percentage of Vulnerable PI Creation
M - Method
%VulWr - Percentage of Vulnerable PI Transfer (WrappingIntent)
%TotVulPI - Percentage of Total PI Vulnerability
ProBr - Protected Broadcast
%ProBr - Percentage of Protected Broadcast
TotBr - Total Broadcast
%TotVulBr - Percentage of Total Broadcast Vulnerability

Figure 6.3 Empirical study with CICMalDroid-2020, Drebin, and Google Play Store datasets

Store [39] apps with App categories such as Productivity, Medical, Parenting; and 10,000,000+ installs. From the dataset, we extract the following statistical information: (1) PendingIntent flags; (2) usage of public broadcast; (3) usage of WrappingIntent; and (4) finally, vulnerable PendingIntent creation statistics (displayed in Figure 6.3).**Binary analysis:** This section explains our tool called Murax, which we have developed to extract the PI information from .apk files. The architecture of Murax is shown in Figure 6.4.

First, Murax pre-processes the .apk files by converting the .dex file (Dalvik Executable) into a .smali file (a human-readable format for the DEX binary code [35] by using open source libraries like Baksmali [34] and Dexlib2 [36]. Next, our Smali parser (written in ANTLR/Java) parses the generated .smali files by extracting the required information like PI flags, PI

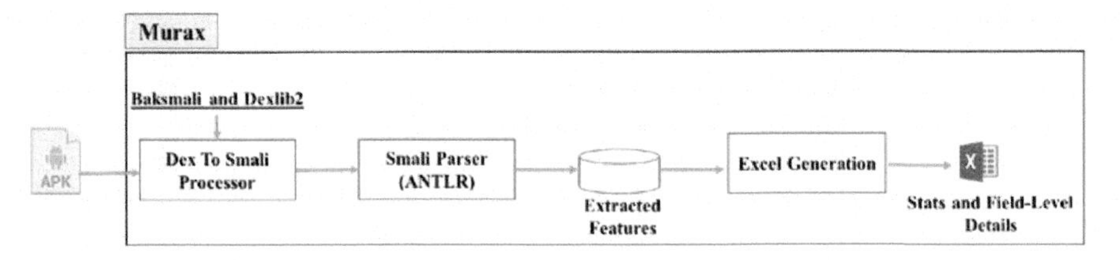

Figure 6.4 Data extraction architecture

creations, and PI transfer. Finally, our Excel Generation module organizes the extracted information as an excel file. Figure 6.3 shows our findings from the dataset.

Explanation: Figure 6.3 displays our empirical study results from 23,922 apps, i.e., a collection of ~20 million classes composed of ~97 million methods. In our analysis, we have classified the vulnerabilities in PendingIntent under two categories: (1) VulPI – vulnerabilities during PI creation (i.e., when the base intent of the PI is either empty or uses public broadcast actions); (2) VulWr – exposing vulnerable PI through public broadcast.

The PI vulnerability percentage calculation is given as follows:

$$\%\text{VulPI} = \frac{\#\text{VulPI}}{\#\text{PI}} * 100 \tag{6.1}$$

$$\%\text{VulWr} = \frac{\#\text{VulWr}}{\#\text{PubBr}} * 100 \tag{6.2}$$

$$\%\text{ProBr} = \frac{\#\text{ProBr}}{\#\text{PubBr}} * 100 \tag{6.3}$$

$$\%\text{TotalVulnerability} = \%\text{VulPI} + \%\text{VulWr} + \%\text{ProBr}. \tag{6.4}$$

Some of our interesting findings from the dataset are given as follows:

1. The presence of ~14\% PI vulnerabilities in the benign tagged apps from [37].
2. The presence of ~10\% PI vulnerabilities in Google Play Store apps.
3. Figure 6.5 shows the statistics on the usage of unsafe PI and protected broadcast in our dataset.
4. Figure 6.6 shows the statistics on the total vulnerability percentage for broadcast and PendingIntent.

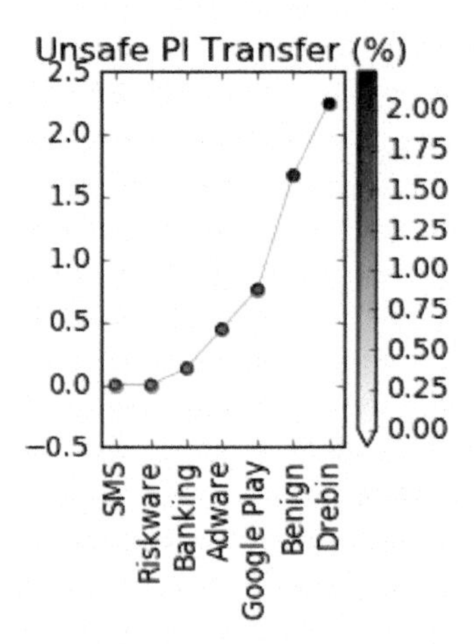

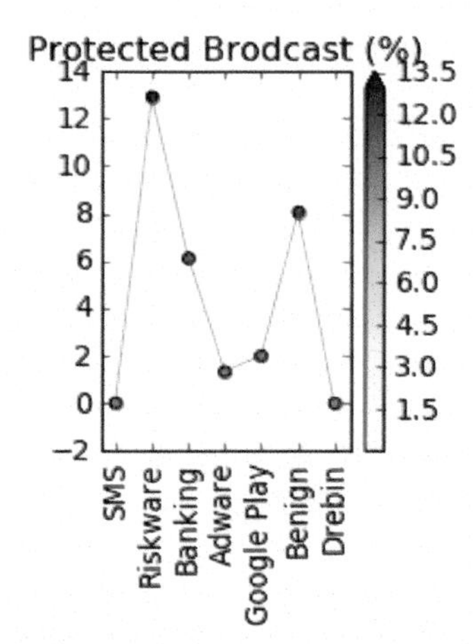

Figure 6.5 Percentage of unsafe PI (low value is better) and protected broadcast (high value is better)

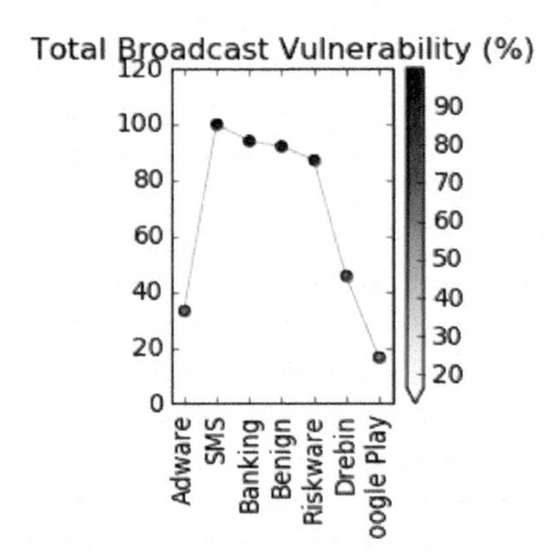

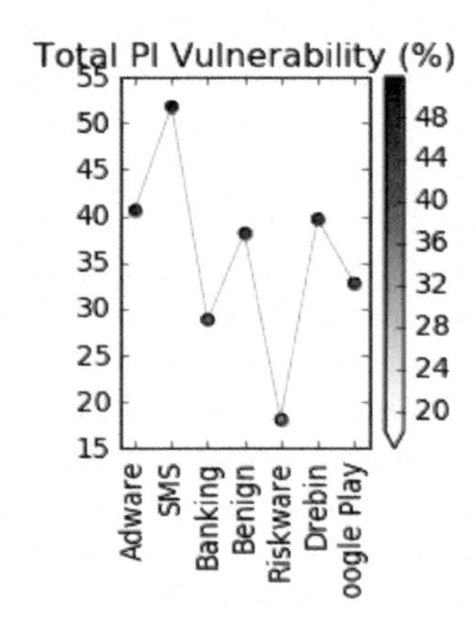

Figure 6.6 Total broadcast vulnerabilities and PI vulnerabilities (low value is better)

5. "FLAG_UPDATE_CURRENT" is the most frequently used PI flag (i.e., ∼53\%), and "FLAG_IMMUTABLE" is one of the sparsely used flags (i.e., ∼1\%) during PendingIntent creation.
6. ∼.21\% of apps have a vulnerable PI transfer (though it looks trivial, these vulnerabilities can create precarious behavior in apps).
7. In total, ∼37\% of apps have PI vulnerabilities.
8. In total, ∼66\% of apps have broadcast vulnerabilities (i.e., usage of implicit broadcast instead of either protected or explicit broadcast).

noOfClz	noOfMeth	noOfInten	noOfPendi	FLAG_ONE	FLAG_NO_	FLAG_CAN	FLAG_UPD	FLAG_IMM	unsafePICr	unsafeBro	unsafePITr	NoOfProte	UnsafePro
187	1087	26	2	0	0	0	0	0	0	2	0	0	0
37	206	13	2	0	0	2	0	0	0	0	0	0	0
176	1038	59	6	0	0	0	0	0	0	4	0	0	0
1087	7045	188	25	0	0	5	19	0	0	6	0	0	0
1314	7203	401	30	0	0	6	10	5	5	15	0	0	0
198	1116	30	2	0	0	0	0	0	0	2	0	0	0
125	934	52	6	0	0	3	0	0	1	3	0	0	0
443	2264	96	30	1	0	0	14	0	2	10	0	0	0
1029	4681	102	6	0	0	4	1	0	0	1	0	0	0
405	2511	55	6	0	0	5	0	0	0	4	0	0	0

Figure 6.7 Dataset snippet (displaying 10 examples)

6.3.2 Dataset

The dataset consists of 25 attributes and 23,277 samples. Each sample (or record) describes the properties like (1) number of classes and methods present in an application (Column 1:2); (2) number of intents and PendingIntents present in an application (Column 3:4); (3) the PendingIntent (PI) flags that are used during the PI creation (Column 5:9); (4) number of protected broadcast and broadcast-related inference (Column 10:14); (5) number of shared user IDs (shareduid) used (Column 15); (6) justifying the intent and PendingIntent leaks (Column 16:23); (7) PI-based error classification (five classes) (Column 24); (8) the final expected prediction classification (Column 25).

The dataset snippet is given in Figure 6.7, where every attribute value is represented as an integer number.

6.3.3 Random Oversampling and Outlier Pre-processing

Imbalanced datasets are those with a severe skew in the class distribution, such as a ratio of 1:100 or 1:1000 data samples from the minority class to the majority class. This skewness in the training dataset can influence many machine learning algorithms by introducing bias in the prediction and thereby ignoring the minority class entirely. Thus, the information carried out by the minority class is ignored during the learning process. The pre-processing method applies specific random resampling techniques to avoid such imbalanced dataset. There are two generally known techniques: (1) undersampling – where the records from the majority class will be deleted to even the class ratio; (2) oversampling – where the minority class records will be duplicated and thereby increase the ratio between minority to majority classes.

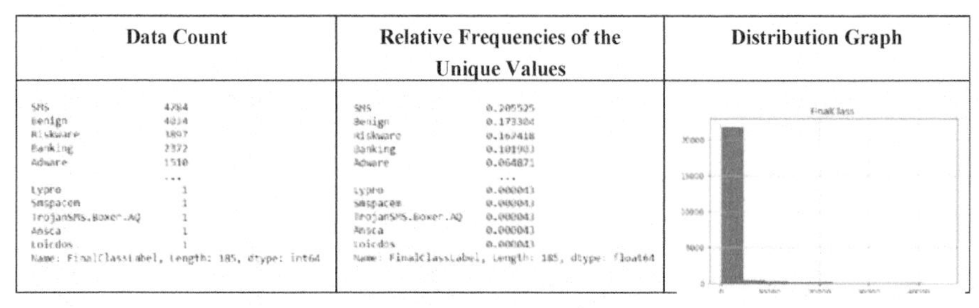

Data Count	Relative Frequencies of the Unique Values	Distribution Graph
SMS 4784 Benign 4034 Riskware 3897 Banking 2372 Adware 1510 ... Lypro 1 Smspacem 1 TrojanSMS.Boxer.AQ 1 Ansca 1 Loicdos 1 Name: FinalClassLabel, Length: 185, dtype: int64	SMS 0.205525 Benign 0.173304 Riskware 0.167418 Banking 0.101903 Adware 0.064871 ... Lypro 0.000043 Smspacem 0.000043 TrojanSMS.Boxer.AQ 0.000043 Ansca 0.000043 Loicdos 0.000043 Name: FinalClassLabel, Length: 185, dtype: float64	

Table 6.1 Imbalanced data

In this paper, we apply the random oversampling technique (a method of random resampling). Random oversampling duplicates records from the minority class in the training dataset and adds them to the original training dataset. SMOTE is another oversampling algorithm that relies on nearest neighbors to create its synthetic data [40]. The comparison of random oversampling and SMOTE is given below:

1. It can result in overfitting for some machine learning models.
2. It increases the size of the training dataset through repetition of the original records, and it does not cause any increase in the variety of training records.
3. On the other hand, oversampling using SMOTE increases the size of the training dataset and also increases the variety in the minority data.
4. Though SMOTE generates an oversampled dataset with variety, it has a limitation when the minority example count equals one (i.e., $n = 1$).

Since our dataset contains few minority classes with only one example, we cannot apply SMOTE. Thus, we prefer to apply the random oversampling model over SMOTE.

Table 6.1 displays our dataset. Columns 1 and 3 show the data count with an imbalance in distribution. Column 2 shows the corresponding relative frequencies of the data. This shows the requirement for an oversampling technique to balance the dataset for training.

6.3.4 Correlation Calculation

The app repository classifies the apps into high-level classes like (1) the app into Multicollinearity occurs when there are two or more independent variables in a multiple regression model, which have a high correlation among themselves. When there are some highly correlated features, we might have difficulty distinguishing between their individual effects on the dependent variable. Multicollinearity can be detected using various techniques: the variance inflation factor (VIF). The association amongst the attributes and representation of a linear relationship is demarcated using correlation. The correlation lies between −1 and +1, where −1, 0, and +1 indicate the perfect negative correlation, no correlation, and perfect positive correlation. The coefficient of correlation can be calibrated mathematically using the following formula:

$$\mathrm{r} = \frac{\mathrm{n}\sum(\mathrm{xy}) - (\sum \mathrm{x})(\sum \mathrm{y})}{\sqrt{\mathrm{n}\sum \mathrm{x}^2 - (\sum \mathrm{x})^2}\sqrt{\mathrm{n}\sum \mathrm{y}^2 - (\sum \mathrm{y})^2}} \tag{6.5}$$

Figure 6.8 displays the correlation between attributes. It shows a high correlation between CombinedClass and FinalClass attributes. Similarly, the other correlations are as follows:

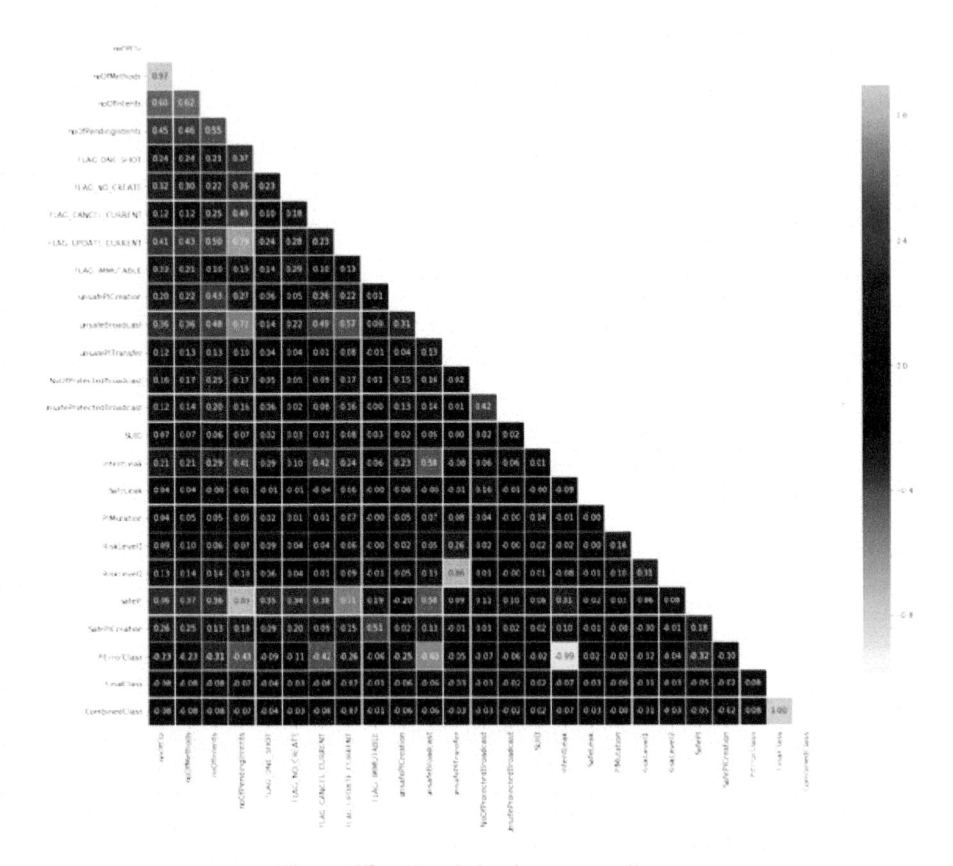

Figure 6.8 Correlation between attributes

1. RiskLevel2 and unsafeBroadcast correlate 0.86.
2. SafePI and noOfPendingIntents correlate 0.89.
3. noOfMethods and noOfClz correlate 0.97.

The correlation coefficients between the variables give us the extent of multicollinearity present in the dataset. For the above-correlated attributes, we calculate the VIF and tolerance values. The variance inflation factor (VIF) and tolerance are closely related statistics for diagnosing collinearity in multiple regression. VIF measures how much the variance of an estimated regression coefficient increases if your predictors are correlated. The higher the value of VIF for the attribute, the more it is highly correlated to other variables. In our case, we choose a threshold of 5 which means that if VIF is more than 5 for a particular variable, then that variable will be removed.

Calculating variance inflation factor (VIF):

- If VIF = 1, no multicollinearity between attributes
- If VIF $\leq$ 5, low multicollinearity or moderately correlated data
- If VIF $\geq$ 5, high multicollinearity or highly correlated data

Tolerance is the level of error that is acceptable for an engineering application. Tolerance is the reciprocal of VIF.

Calculating tolerance (reciprocal of VIF):

- If VIF is high, then tolerance will be low (i.e., high multicollinearity).
- If VIF is low, the tolerance will be high (i.e., low multicollinearity).

In Table 6.2, we calculate the VIF and tolerance between the attributes CombinedClass and FinalClass. Since the value is more than our threshold value (5), we remove either one of the columns and recalculate the VIF and tolerance. Thus, we can find a resulting dataset with low multicollinearity.

	VIF	Tolerance
IntentLeak	1.251168e+02	7.992530e-03
RiskLevel1	1.106252e+00	9.039534e-01
RiskLevel2	3.007710e+00	3.324788e-01
CombinedClass	9.995801e+08	1.000420e-09
FinalClass	9.995281e+08	1.000472e-09

	VIF	Tolerance
IntentLeak	1.011787	0.988351
RiskLevel1	1.105841	0.904290
RiskLevel2	1.113403	0.898148
FinalClass	1.006141	0.993896

	VIF	Tolerance
IntentLeak	1.011838	0.988300
RiskLevel1	1.105841	0.904290
RiskLevel2	1.113406	0.898145
CombinedClass	1.006194	0.993844

Table 6.2 VIF and tolerance between attributes

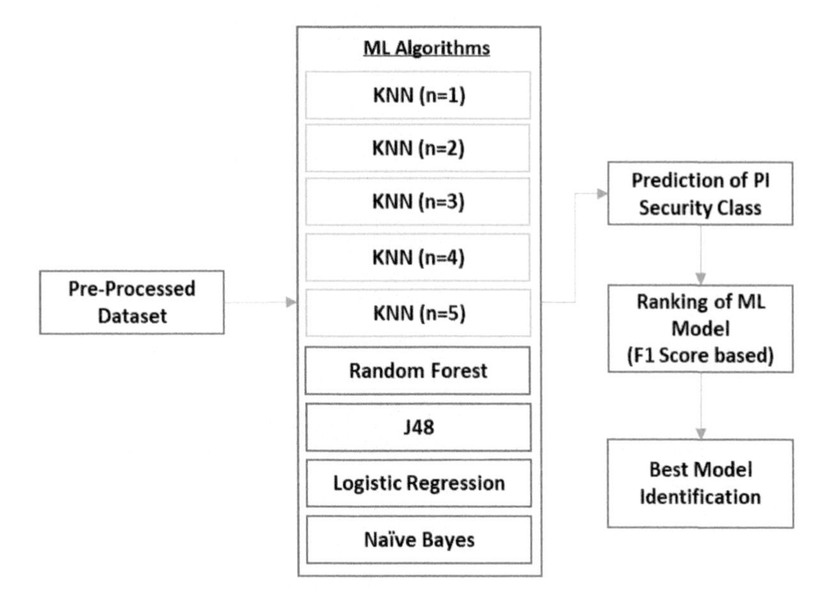

Figure 6.9 Selecting the best ML model

6.4 Identification of Best Machine Learning Model

This section explains various machine learning models that we have applied to test our dataset. Figure 6.9 shows the machine learning models that we have used to test the dataset.

6.4.1 Confusion Matrix

The confusion matrix was used to analyze the performance measurement of the imbalanced classification (Table 6.3). True positive (TP), true negative (TN), false positive (FP), and false negative (FN) were the four distinct combinations of predicted value and real value of the confusion matrix. The accuracy, precision, recall, and $F1$ scores could be enumerated by this method.

6.4.2 Accuracy

Accuracy is the degree of all the precisely identified scenarios. It is preferred when all the attributes are considered equal and if the class distribution is balanced.

Table 6.3 The confusion matrix

		Actual Values	
		Positive (1)	Negative (0)
Predicted Values	Positive (1)	True positive (TP)	False positive (FP)
	Negative (0)	False negative (FN)	True negative (TN)

$$\text{Accuracy} = \frac{\text{TP} + \text{TN}}{\text{TP} + \text{TN} + \text{FP} + \text{FN}}. \tag{6.6}$$

6.4.3 Precision

The precision determines how many are positive among the predicted values from the overall positive classes. The precision value is anticipated when the number of false positives is minimal.

$$\text{Precision} = \frac{\text{TP}}{\text{TP} + \text{FP}}. \tag{6.7}$$

6.4.4 Recall

The recall finds the accurately predicted values, from all the positive classes. It is also called true positive rate (TPR) or sensitivity. The recall value is much more desirable when the number of false negatives is minimal.

$$\text{Recall} = \frac{\text{TP}}{\text{TP} + \text{FN}}. \tag{6.8}$$

6.4.5 F1Score

$F1$ score or $F1$ measure is the equilibrium between precision and recall. It is considered the most effective metric to quantify real-world problems since it mostly contains imbalanced class distribution.

$$F1\ \text{Score} = 2 \times \left[\frac{\text{Precision} \times \text{Recall}}{\text{Precision} + \text{Recall}}\right]. \tag{6.9}$$

6.4.6 AUC-ROC

The performance of the machine learning models could also be measured by receiver operating characteristics (ROC) and area under the curve (AUC).

The capability of the classification model would be defined by the ability to disjuncture the classes. The ROC curve would be formed by plotting the false positive rate (FPR) and true positive rate (TPR) in the x- and y-axes, respectively. Higher the prediction of true positive (TP) and true negative (TN) values by model results with elevated AUC rate. The AUC would lie between 0 and 1. When the AUC move toward "0," it represents the worst prowess of disjuncture; while it progresses toward "1," it signifies the best proficiency of disjuncture. Finally, the machine learning models are ranked based on their $F1$ score to identify the PI-based security. The $F1$ score combines the precision and recall into a single metric by taking their harmonic mean. Since the $F1$ score is the weighted average (or harmonic mean) of precision and recall, it takes both false positives and false negatives into account to balance the precision and recall of the classifier.

For example, the accuracy percentage (%) can be biased by many true negatives, when the data has a high asymmetry which results in a very high accuracy score. That is, accuracy assesses the classifier's competence at identifying both true positives and true negatives rather than assessing the classifier's competence based on its capability to identify the true positives alone. Thus, our final decision for the best model is ranked based on the $F1$ score rather than accuracy.

6.5 Discussion

The Naïve Bayes (NB), logistic regression (LR), random forest (RF), K-nearest neighbor (KNN), and J48 algorithms were employed to predict the PI-based security attacks.

Table 6.4 displays the metrics of the classifier we have tested. Figure 6.10 displays the corresponding ROC curve of Table 6.4 data, i.e., data before applying RandomSampling. We have used "micro" averaging for our test, which results in the same value for all the four metrics (accuracy, precision, recall, and $F1$ score). The result shows that micro averaging does not distinguish between different classes, and, thus, it averages their metric scores rather than providing weightage based on the label. This shows that the dataset has an unequally distributed test set (e.g., 178 classes, which has 90% of the samples).

AUC-ROC curve is a performance measurement for the classification problems at various threshold settings. ROC is a probability curve and AUC represents the degree or measure of separability; thus, it specifies the model's capability to distinguish the classes (shown in Figure 6.10). With a higher

Table 6.4 The outcomes of the algorithms (before RandomSampling)(type of averaging performed on the data = "micro")

Algorithm	Accuracy (%)	Precision	Recall	*F*1 score	ROC area
Naïve Bayes	20.3	0.203	0.203	0.203	0.435
Logistic regression	40.2	0.402	0.402	0.402	0.229
KNN (n = 1)	70.3	0.703	0.703	0.703	0.486
KNN (n = 2)	70.4	0.704	0.704	0.704	0.485
KNN (n = 3)	69.8	0.698	0.698	0.698	0.482
KNN (n = 4)	70.0	0.700	0.700	0.700	0.479
KNN (n = 5)	68.5	0.685	0.685	0.685	0.483
J48	74.9	0.749	0.749	0.749	0.469
Random forest	78.7	0.787	0.787	0.787	0.582

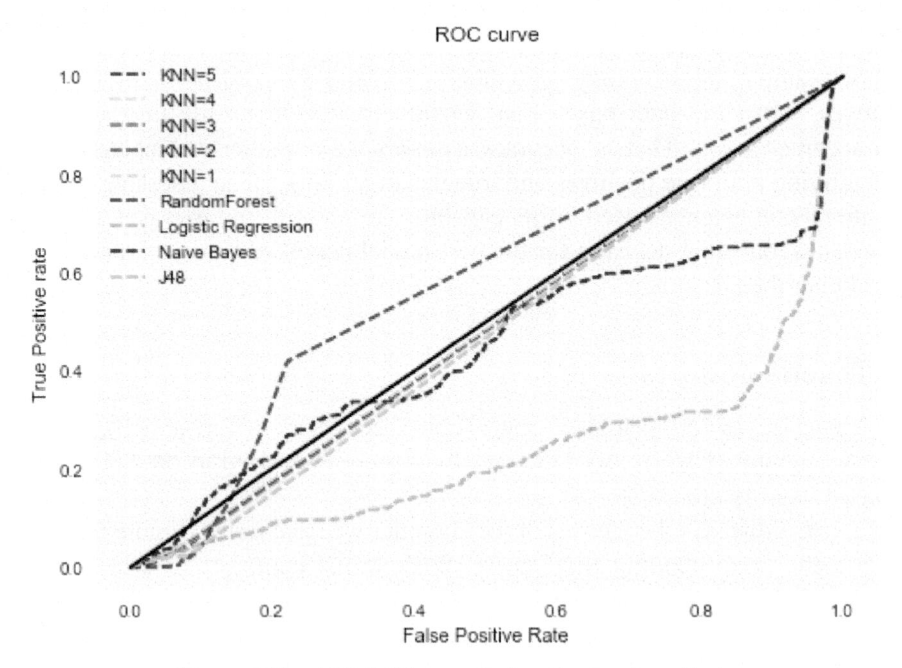

Figure 6.10 AUC-ROC curve (before RandomSampling)

AUC value, the model better predicts the classes correctly. A good model has AUC near 1, which means it has a good separability measure. When a model is poor in classification, it has an AUC near 0, which means that it has the worst separability measure. When AUC is 0.5, the model has no class separation capacity.

Table 6.5 and Figure 6.11 display the analysis after RandomSampling. From Table 6.5, we can observe that the ROC value of random forest improved slightly which means the algorithm after RandomSampling performs better at ranking the test data, with most negative cases at one end of a scale and positive cases at the other.

Table 6.5 The outcomes of the algorithms (after RandomSampling)(type of averaging performed on the data = "micro")

Algorithm	Accuracy (%)	Precision	Recall	*F*1 score	ROC area
Logistic regression	12.9	0.129	0.129	0.129	0.257
Naïve Bayes	14.7	0.147	0.147	0.147	0.464
KNN (n = 1)	70.3	0.703	0.703	0.703	0.486
KNN (n = 2)	71.2	0.712	0.712	0.712	0.485
KNN (n = 3)	68.9	0.689	0.689	0.689	0.486
KNN (n = 4)	70.0	0.700	0.700	0.700	0.490
KNN (n = 5)	64.3	0.643	0.643	0.643	0.490
J48	69.6	0.696	0.696	0.696	0.472
Random forest	73.8	0.738	0.738	0.738	0.640

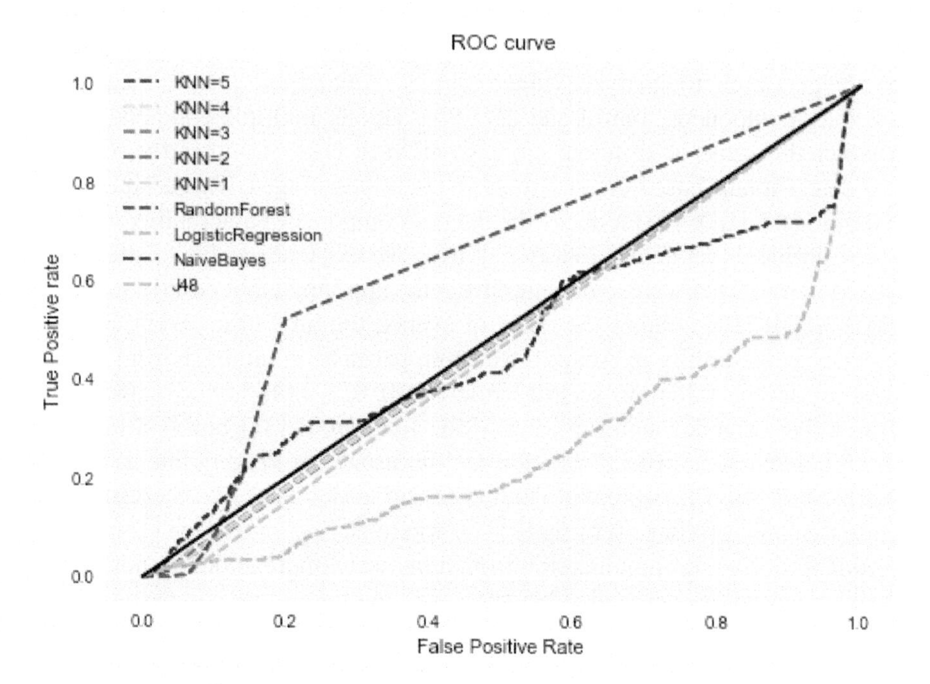

Figure 6.11 AUC-ROC curve (after RandomSampling)

6.6 Related Work

Detecting data leaks and malicious components during ICC have already been explored for source code and DEX files. These approaches combine static analysis, dynamic analysis, and runtime policies at the OS level by modifying the Android framework or kernel code. However, PendingIntent is designed and implemented as a pluggable Java library that dynamically filters the data leaks and malicious components during ICC. RAICC [20] and PIAnalyzer [27] study security threats on PendingIntent. RAICC instruments the PendingIntent calls (AICC calls) by adding a method startActivity() with the right intent as a parameter, thereby converting them to standard ICC calls that can be further processed with IccTA [21] and Amandroid [18]. RAICC extends DroidBench [28] with 20 PendingIntent benchmarks, enables IccTA and Amandroid to handle PendingIntents, and improves the precision of IccTA by 88.90% and Amandroid by 83.33%.

Static analysis and dynamic monitoring help to prevent ICC attacks [17], [28], [32]. XmanDroid [29] is the policy-based model available in the literature. SALMA [24]uses an incremental machine learning model to analyze app security. Barros *et al.* [25] use security annotation on the intent data to track the taint flow. Incremental machine learning model [12], [24] builds a knowledge graph to analyze vulnerabilities. Amandroid [18] statically detects inter-component control and data flow among multiple components from either the same or different apps. ComDroid [16] examines the .dex files to detect intent-based attacks such as intent spoofing, UIR, and intent vulnerabilities. CHEX [23] reduces the data or component vulnerability as a data-flow problem that helps identify risk from the existence of flow pattern. AnFlo [11] groups malware and benign from the app functional descriptions. TERMINATOR [10] collates the order of events, the time of an event, and the security vulnerability in Android's custom permission model. Kirin [30] certifies an app install time security rules. DREBIN [14] infers malware patterns rather than manually crafting [30]. TaintDroid [31] builds a taint graph on context-sensitive information and its usage. In PendingIntent, we do not provide such fine-grained context-sensitive policies. An interesting paper on dynamic privilege scoring is [17], where an app's privilege score is reduced if it receives communication from a malware application. DroidCap [8] facilitates privilege separation between app components by compartmentalizing components into a logical app with a subset of privileges. Maxoid [33] confines the delegates from leaking secrets by creating different views of the initiator's state. Aquifer [22] defines UI workflow policies where each participating application defines its export list, a required list, and a workflow filter. Mutent [26] is the first ownership-based [13], [15], [19] method to protect implicit intent from malware attacks.

When data is exchanged through protected or unprotected channels, the system must follow the encryption technique [41]. Geetha *et al.* [42] apply machine learning to security analysis and performance improvement. Maheswari *et al.* [43] use a genetic algorithm to classify the IDS system. Sindhu *et al.* [44] classify the app features by applying probabilistic matrix factorization (PMF) with version evolution progress model (VEPM), thereby considering the app version (a timeline feature) into the app recommendation. A python-based record and play environment called Culebra [5] can be helpful to test an app and collect data. And for testing purposes, we have used some Github-based open-source apps [1]–[4]. These four apps were modified with PI exchange and tested for accuracy.

6.6.1 Limitations and Future Work

Currently, the model has the following limitations in training, say, categorization of apps based on other criteria like app permissions, app creators, etc. We will try to add the app permissions and creator ratings to train the model in the future.

6.7 Conclusion

According to our knowledge, the literature has limited research works on PendingIntent security. We have presented a mechanism to analyze the APK for PendingIntent-based security issues automatically. By building an ML model, we quickly capture the PI security issues. The proposed ML-based model is the first holistic self-learning framework by classifying the app security based on the PendingIntent properties.

References

[1] Android-Alarm, https://github.com/leanh153/Android-Alarm

[2] Android Custom Push Notification Layouts, https://github.com/WebEngage/android-custom-push-layouts

[3] Android Keynotes, https://github.com/akash2099/KeepNotes-Android App

[4] Android Notification, https://github.com/jaisonfdo/NotificationExample

[5] Culebra: Generating Ready-to-Execute Scripts for Black Box Testing, https://github.com/dtmilano/AndroidViewClient/wiki/culebra

[6] PendingIntent, https://developer.android.com/reference/android/app/PendingIntent

[7] sendBroadcastWithMultiplePermissions, https://developerandroid.com/reference/android/content/Context#sendBroadcastWithMultiplePermissions(android.content.Intent,java.lang.String[])

[8] Abdallah Dawoud, Sven Bugiel: DroidCap: OS Support for Capability-Based Permissions in Android. In: NDSS Symposium (2019).

[9] Ali Feizollah, Nor Badrul Anuar, Rosli Salleh, Guillermo Suarez-Tangil, Steven Furnell: AndroDialysis: Analysis of Android Intent Effectiveness in Malware Detection. In: Elsevier Computers & Security. pp. 121–134 (2017).

[10] Alireza Sadeghi, Reyhaneh Jabbarvand, Negar Ghorbani, Hamid Bagheri, Sam Malek: A Temporal Permission Analysis and Enforcement Framework for Android. In: 40th ICSE. pp. 846–857 (2018).

[11] Biniam Fisseha Demissie, Mariano Ceccato, Lwin Khin Shar: AnFlo: Detecting Anomalous Sensitive Information Flows in Android Apps. In: MOBILESoft. pp. 24–34 (2018).

[12] Biniam Fisseha Demissie, Mariano Ceccato, Lwin Khin Shar: Security Analysis of Permission Re-Delegation Vulnerabilities in Android Apps, Empirical Software Engineering, volume 25, pp. 5084–5136, 2020.

[13] David G. Clarke, James Noble, John M. Potter: Simple Ownership Types for Object Containment. In: ECOOP 2001. LNCS, volume 2072. Springer, Berlin (2001).

[14] Daniel Arp, Michael Spreitzenbarth, Malte Hubner, Hugo Gascon, Konrad Rieck: DREBIN: Effective and Explainable Detection of Android Malware in Your Pocket. In: NDSS'14. pp. 23–26 (2014).

[15] David G. Clarke, Sophia Drossopoulou: Ownership, Encapsulation and the Disjointness of Type and Effect. In: 2002 ACM OOPSLA 2002. ACM, pp. 292–310 (2002).

[16] Erika Chin, Adrienne Porter, Felt Kate Greenwood, David Wagner: Analyzing Inter-Application Communication in Android. In: 9th MobiSys. pp. 239–252 (2011).

[17] Adrienne Porter Felt, Helen J. Wang, Alexander Moshchuk, Steve Hanna, Erika Chin: Permission Re-Delegation: Attacks and Defenses. In: USENIX Symposium. p. 22 (2011).

[18] Fengguo Wei, Sankardas Roy, Xinming Ou, Robby: Amandroid: A Precise and General Inter-Component Data Flow Analysis Framework for Security Vetting of Android Apps. In: ACM CCS'14. pp. 1329–1341 (2014).
[19] John Potter, James Noble, Dave Clarke: The ins and outs of objects. In: 1998 Australian Software Engineering Conference. IEEE Computer Society, Washington, DC, USA, pp. 80–89 (1998).
[20] Jordan Samhi, Alexandre Bartel, Tegawendé F. Bissyandé, Jacques Klein: RAICC: Revealing Atypical Inter-Component Communication in Android Apps. In: 43rd International Conference on Software Engineering (ICSE). IEEE/ACM, Madrid, Spain (2021).
[21] Li Li, Alexandre Bartel, Bissyandé, Tegawendé F. Bissyandé, Jacques Klein, Yves Le Traon, Steven Arzt, Siegfried Rasthofer, Eric Bodden, Damien Octeau, Patrick McDaniel: IccTA: Detecting Inter-Component Privacy Leaks in Android Apps. In: 2015 IEEE/ACM 37th IEEE International Conference on Software Engineering. volume 1, pp. 280–291 (2015).
[22] Limin Jia, Jassim Aljuraidan, Elli Fragkaki, Lujo Bauer, Michael Stroucken, Kazuhide Fukushima, Shinsaku Kiyomoto, Yutaka Miyake: Run-Time Enforcement of Information-Flow Properties on Android. In: European Symposium on Research in Computer Security. pp. 775–792 (2013).
[23] Long Lu, Zhichun Li, Zhenyu Wu, Wenke Lee, Guofei Jiang: CHEX: Statically Vetting Android Apps for Component Hijacking Vulnerabilities. In: ACM CCS. pp. 229–240 (2018).
[24] Mahmoud Hammad, Joshua Garcia, Sam Malek: Self-Protection of Android Systems from Inter-Component Communication Attacks. In: 33rd ACM/IEEE ASE'18. pp. 726–737 (2018).
[25] Paulo Barros, René Just, Suzanne Millstein, Paul Vines, Werner Dietl, Marcelo d'Amorim, Michael D. Ernst: Static Analysis of Implicit Control flow: Resolving Java Reflection and Android Intents. In: 30th ASE'15. pp. 669–679 (2015).
[26] Pradeep Kumar D.S., Tiffany Bao, Jaejong Baek, Yan Shoshitaishvili, Adam Doupé, Ruoyu Wang, Gail-Joon Ahn: MuTent: Dynamic Android Intent Protection with Ownership-Based Key Distribution and Security Contracts. In: 54th HICSS'54. pp. 7217–7226 (2021).
[27] Sascha Groß, Abhishek Tiwari, Christian Hammer: PIAnalyzer: A Precise Approach for PendingIntent Vulnerability Analysis. In: European

Symposium on Research in Computer Security. Springer, pp. 41–59 (2018).

[28] Steven Arzt, Siegfried Rasthofer, Christian Fritz, Eric Bodden, Alexandre Bartel, Jacques Klein, Yves Le Traon, Damien Octeau, Patrick McDaniel: FlowDroid: Precise Context, Flow, Field, Object-Sensitive and Lifecycle-Aware Taint Analysis for Android Apps. In: PLDI'14. pp. 259–269 (2014).

[29] Sven Bugiel, Lucas Davi, Alexandra Dmitrienko, Thomas Fischer, Ahmad-Reza Sadeghi: XManDroid: A New Android Evolution to Mitigate Privilege Escalation Attacks. In: Technische Universität Darmstadt, TR-2011-04 (2011).

[30] William Enck, Machigar Ongtang, Patrick McDaniel: On Lightweight Mobile Phone Application Certification. In: ACM CCS. pp. 235–245 (2009).

[31] William Enck, Peter Gilbert, Byung-Gon Chun, Landon P. Cox, Jaeyeon Jung, Patrick Mcdaniel, Anmol N. Sheth: TaintDroid: An Information-Flow Tracking System for Realtime Privacy Monitoring on Smartphones. In: 9th USENIX (OSDI'10). pp. 393–407 (2010).

[32] Youn Kyu Lee, Jae Bang, Gholamreza Safi, Arman Shahbazian, Yixue Zhao, Nenad Medvidovic: A SEALANT for Inter-app Security Holes in Android. In: 39th ICSE. pp. 312–323 (2017).

[33] Yuanzhong Xu, Emmett Witchel: Maxoid: Transparently Confining Mobile Applications with Custom Views of State. In: EuroSys'15. pp. 1–16 (2015).

[34] Baksmali, https://github.com/JesusFreke/smali/tree/master/baksmali, Last-Visited-(08/1/2022).

[35] DexLib2, https://github.com/JesusFreke/smali/tree/master/dexlib2, Last-Visited-(08/01/2022).

[36] Smali, https://github.com/JesusFreke/smali,Last-Visited-(08/01/2022)

[37] CICMalDroid 2020, https://www.unb.ca/cic/datasets/maldroid-2020.html

[38] The Drebin Dataset, https://www.sec.cs.tu-bs.de/~danarp/drebin/index.html

[39] Google Play, https://play.google.com/store

[40] Nitesh V. Chawla, Kevin W. Bowyer, Lawrence O. Hall, W. Philip Kegelmeyer, SMOTE: Synthetic Minority Over-Sampling Technique, Journal of Artificial Intelligence Research, volume 16, 321–357, 2002.

[41] Geetha Subbiah, P. Punithavathi, A. Magnus Infanteena, S. Siva Sivatha Sindhu: A Literature Review on Image Encryption Techniques, International Journal of Information Security and Privacy (IJISP), 2018.

[42] Geetha Subbiah, S. Siva Sivatha Sindhu, Nagappan Kamaraj: Blind Image Steganalysis Based on Content Independent Statistical Measures Maximizing the Specificity and Sensitivity of the System, Computers & Security, volume 28 (7), 683–697, 2009.

[43] M. Maheswari, Geetha Subbiah, S. Selva Kumar, Marimuthu Karuppiah, Debabrata Samanta, Yohan Park: PEVRM: Probabilistic Evolution Based Version Recommendation Model for Mobile Applications, IEEE Access, volume 9, 20819–20827, 2021.

[44] S. Siva Sivatha Sindhu, Geetha Subbiah, M. Marikannan, Kannan Arputharaj: A Neuro-Genetic Based Short-Term Forecasting Framework for Network Intrusion Prediction System, International Journal of Automation and Computing, volume 6 (4), 406, 2009.

7

Machine Learning and Blockchain Integration for Security Applications

Aradhita Bhandari[1], Aswani Kumar Cherukuri[1], and Firuz Kamalov[2]

[1]School of Information Technology and Engineering, Vellore Institute of Technology, India
[2]School of Electrical Engineering, Canadian University Dubai, UAE
E-mail: aradhita02@gmail.com; cherukuri@acm.org

Abstract

The modern world is connected by technology, with endless end devices, servers, infrastructures, and other resources creating a complex and interesting cyber landscape, constantly changing and improving, with new technologies discovered and vulnerabilities uncovered every day. In the last few years, machine learning and Blockchain technology have independently gained robust solutions for many cybersecurity problems. Machine learning has allowed for the development of smarter security systems that can automate processes such as intrusion or anomaly detection, allocation of resources, and scalability of operations. Blockchain technology has opened up more possibilities for decentralized systems that maintain or enhance security and privacy. Further, Blockchain technology has also renewed interest in smart contracts that automatically uphold agreements between parties involved. Unfortunately, their popularity has also exposed new vulnerabilities to other threats. The unique features of Blockchain and machine learning leave room for each to improve the other. Blockchain offers decentralized security and trust for the models and data for machine learning, while machine learning provides intelligent decision-making to anomaly detection, scalability, and efficiency in Blockchain networks. This work highlights how an interplay between Blockchain and ML would allow both technologies to assist cybersecurity-related use cases. In this chapter, cybersecurity is defined, and common vulnerabilities are outlined, following which each of the technologies mentioned above is discussed in detail. Finally, developments involving both technologies are presented, and areas for future research are identified.

Keywords: Cybersecurity, machine learning, Blockchain technologies, smart contracts.

7.1 Introduction

A large part of the modern world exists in the cyber landscape, and much focus is being given to digitizing even the more mundane aspects of our lives. This near ubiquity of technology, despite all of its benefits, also presents a complex, variable, and interesting attack surface for vulnerabilities. As new technologies are developed, including security, more vulnerabilities are inadvertently added. The effort to preserve safety is a constant battle of wits between attackers and researchers. Two major players in cybersecurity in the last few years are machine learning and Blockchain technologies [2].

Blockchain technologies, first introduced in 2008, are a huge leap away from traditional centralized systems [11]. Blockchains introduce the aspects of trust and security to decentralized systems via immutable ledgers and majority consensus amongst various nodes in the network. The technology allowed for creating the first cryptocurrency and bolstered developments in fintech. Since then, its use has expanded from finance to various other domains, with much research on Blockchain-based cybersecurity [55].

Machine learning consists of various algorithms and techniques and can also be considered an umbrella term for artificial intelligence, deep learning, and natural language technologies. Machine learning algorithms attempt to mimic intelligent decision-making processes based on the patterns and connections learned from available data, usually in the frame of one problem. Machine learning has seen various applications, from government defense to medicine and cybersecurity. Despite the constraints often raised by the quality and quantity of data, machine learning has found many uses in security. It is well known for anomaly or intrusion detection by comparing incoming data to previously learned patterns [75].

Despite the significant developments in machine learning and Blockchain for cybersecurity, each technology exposes a new set of vulnerabilities in the system. Blockchain presents a decentralized computational environment that can mitigate the single point of failure and central vulnerability in machine learning systems. Machine learning can imbibe smart decision-making into Blockchain security, contracts, and resource allocation.

The rest of the chapter continues as follows. In Section 7.2, a background is developed in which cybersecurity is discussed in detail, explaining the necessity for this research. Sections 7.3, 7.4, and 7.5 discuss Blockchain technology, smart contracts, and machine learning, respectively. Each of these sections has three subsections: an introduction to the technology, a study of its applications in cybersecurity, and its identified drawbacks in the context of cybersecurity. Section 7.6 discusses developments in cybersecurity that have utilized both Blockchain technology and machine learning, and Section 7.7 offers directions for future research. Finally, the chapter is concluded in Section 7.8.

7.2 Methodology

This work is a survey of the developments in cybersecurity using machine learning and Blockchain technologies in conjunction. The purpose of this chapter was to answer the following questions:

1. How are Blockchain solutions assisting ML-based cybersecurity systems?
2. How are ML solutions assisting Blockchain-based cybersecurity systems?
3. What are some possible directions for future research in ML + BCT cybersecurity solutions?

The literature referenced in this research has been taken primarily from the following databases: IEEExplore, Science Direct, ACM digital library, Elsevier, and the Scopus database. The papers were identified using one or a combination of keywords such as "cybersecurity," "Blockchain," "smart contracts," "machine learning," and "deep learning." Sixty papers were gathered from the initial searches; 12 of these papers did not discuss the domain of cybersecurity and were, therefore, removed from consideration. Papers were filtered based on relevance to the topic; primary interest was given to papers that presented a development, proposed a new system, or highlighted a drawback in either one or a combination of the technologies. Papers that contained repeated content or did not discuss both machine learning and Blockchain solutions were disregarded. After these considerations, 18 papers were selected for the literature review of current topics. Out of these papers,

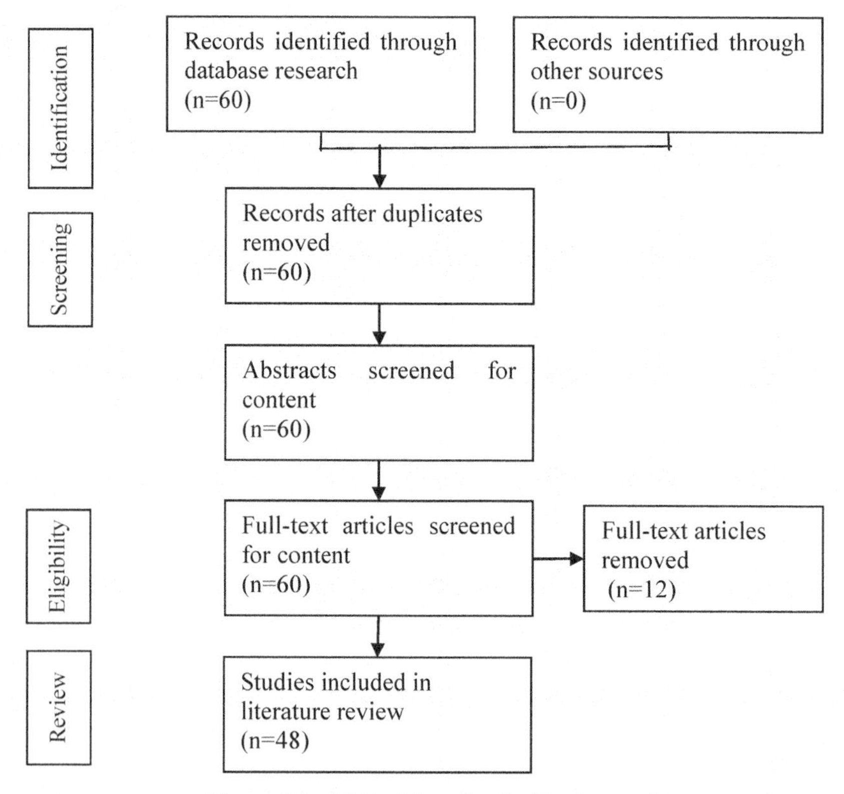

Figure 7.1 Methodology for the literature review

9 discussed ML solutions for Blockchain technologies, and 9 discussed Blockchain solutions for ML technologies in cybersecurity. However, all of the initial 48 papers were considered for concluding possible directions for future research.

7.3 Background

Cybersecurity and information security are often used interchangeably, and while there is considerable overlap, there is also a clear distinction between the two. Many definitions of cybersecurity and information security exist; however, only the International Standards Organization (ISO) definition will be considered.

The ISO defines cybersecurity as *Preservation of confidentiality, integrity and availability of information in the Cyberspace* (ISO/IEC, 2012).

Thus, cybersecurity is information security present over the internet. Blockchain, which requires some sort of network and almost always uses the internet, is primarily discussed in cyberspace.

To understand cybersecurity in depth, three terms in cybersecurity must be defined: confidentiality, integrity, and availability. These, as defined by the ISO:IEC 27000:2018, are as follows.

Confidentiality is the property that information is not made available or disclosed to unauthorized individuals, entities, or processes.

Integrity is the property of accuracy and completeness.

Availability is the property of being accessible and usable on demand by an authorized entity. Cybercrimes are criminal activities that use computer or computer network tools to threaten the confidentiality, integrity, and availability in cyberspace.

Kaur *et al.* [98] identify a broad list of the different kinds of cyberattacks that have been identified. They are split into nine categories: cryptographic attacks, access attacks, reconnaissance attacks, active attacks, passive attacks, phishing attacks, malware attacks, quantum attacks, and web attacks.

Cybercrimes continue to affect individual users, government organizations, and corporations. Even large corporations with considerable investment in their security fall victim to these crimes. Cybercrimes can cause damages or loss of data worth billions of dollars.

From January to March 2021, Microsoft Exchange Servers were vulnerable to attacks by a Chinese state-sponsored hacking group. The group took advantage of four zero-day vulnerabilities to compromise the Outlook Web Access of the servers. The vulnerabilities gave the attackers access to all of the data on the servers, the underlying server networks, and allowed them to escalate their access to administrator privileges. They also installed backdoors on the compromised servers, which gave them access to the servers, even after updates, until they were removed. This attack affected an estimated 250,000 servers [99].

In January 2004, a computer worm named Mydoom was first sighted. It is still used in attacks. It was originally spread via emails, but it managed to take over the internet and affected peer-to-peer (P2P) file-sharing programs. The malware was so prolific that it constituted about a quarter of all emails sent worldwide at its peak. The original version of the malware held two payloads, a backdown to allow control over the subverted PC and a DoS attack against the website of the company SCO group. A modified version of the virus also targeted Microsoft and antivirus websites to forestall antivirus tools and software. The attack led to an estimated loss of $38 billion in damages, making it the most expensive computer malware in history [37].

The Microsoft Exchange attack and Mydoom are two prominent examples of cybersecurity attacks. There are many types of attacks, and each has multiple specific attacks under it. Cybersecurity is a vast and ever-evolving field, and no single cybersecurity solution can be applied to every attack. Ironically, cybersecurity and cyberattacks both support and encourage each other. More often than not, the solutions applied to a particular problem bring a new set of vulnerabilities; a well-known example is that the advent of Blockchain brought with it a new, powerful wave of ransomware. Thus, cybersecurity depends on developing new techniques and technologies to stay ahead of cybercriminals [37].

7.4 Blockchain Technology

7.4.1 Introduction to Blockchain Technology

Nakamoto first introduced the concept of Blockchain in 2008. In the 2008 paper, Nakamoto described Blockchain as an electronic P2P transaction system that could bypass traditional banking solutions [90]. The technology was used to create "bitcoin," a wholly electronic currency, in 2009.

In the Blockchain, the currency is considered a chain of digital signatures that can be used to trace the history of coin ownership. Before transferring the coin, the owner signs the coin by appending a hash of the previous transaction and the public key of the next owner. The main issue that Nakamoto's system addresses is establishing a system of trust that could not be tampered with. It achieved this by using digital signature, hashed timestamps, and the concept of proof-of-work.

Timestamp server openly publishes the hash of a block of items. This timestamp proves the existence of the data in the block at that particular time. Tampering is prevented by including the previous timestamp in the hash of every new timestamp, effectively creating a chain. Implementing the timestamp server without a central authority needs to be distributed over a P2P network.

Proof-of-work is required to implement the timestamp server over the P2P network. The concept was introduced by Back [4], and it defines the average work required as an exponential of the number of zeroes required in the hash, which can be verified by executing a single hash [5]. The proof-of-work is implemented by incrementing a nonce in the block until a value that gives the required zero bits for the hash. Now, CPU effort must be expended to satisfy the proof-of-work, and then the block cannot be changed without redoing the work. Since the blocks form a chain, the older the block is in the chain, the more work is required to modify it, thus providing integrity to the chain.

Proof-of-work also fulfills the purpose of a majority vote. It is essentially one-CPU-one-vote, which is more secure than one-IP-address-one-vote, which can be subverted by anyone with access to multiple IP addresses. If most of the CPU power is controlled by honest nodes, the honest node will grow the fastest and outpace competing chains. Further, the proof-of-work is calculated in inverse proportion to the average number of blocks per hour, which means difficulty increases if blocks are added too fast.

Incentive to work and work as an honest node is included in the system by treating each chain of transactions as a currency. The CPU time and electricity expended are the investment to generate and circulate new coins. Apart from including a balance of currency in the system, this incentive encourages honesty in the nodes. If an attacker were to have access to a lot of CPU power, they are expected to find more return on investment in working honestly than attempting to modify the Blockchain to steal their payments back.

The working of the Blockchain can be described in the following steps:

1. Every new transaction is broadcast to all other nodes with a timestamp.
    (a) As long as a majority of the nodes receive the transaction broadcast, it is alright if some nodes do not. The majority of CPU power would be enough to get the transaction into a block.
2. Each node collects the broadcasted transactions into a block.
3. Each node works to find a proof-of-work for its block.
4. When a proof-of-work is found, the node broadcasts the corresponding block.
5. When a node receives a block, it first verifies that all the transactions in the block are valid.
    (a) If a node in the network does not receive a block for any reason, it will realize as soon as it gets the next block that one is missing. It can then request for the missing block from its peers.
6. If a node accepts the validity of a block, it starts working on creating the next block in the chain. The hash of the previous block will be considered the accepted hash.

(a) Since a longer chain indicates a majority vote, nodes will always consider the longest chain correct and will work on extending it.
(b) If a node receives two blocks of similar chain length, it will work on the first one it receives but will store the other one. The tie between the two blocks will be broken when the next proof-of-work is found.

7.4.2 Applications of Blockchain Technology

Blockchain ended up having a bigger impact than merely introducing an online currency exchange ecosystem. As research soon found, Nakamoto created a secure and scalable structure applicable to various fields other than finance. Blockchain changed the security dynamic in networks, bringing a lot of attention and scope for research.

In traditional models, privacy is achieved by limiting the access to information to the parties involved and the trusted third party. Blockchain bypasses the need for a trusted third party and instead achieves the trust by publicly making the transactions' ledger available. To maintain privacy, the identities of the parties involved in the transaction are hidden [11].

Further, the trusted authorities in traditional systems themselves are vulnerable to attacks. They provide a clear target to attackers. If the trusted authority is compromised in centralized systems, the entire system is compromised [3]. Blockchain avoids this vulnerability by distributing the system, as it is much more difficult to overtake the majority of CPUs in a network simultaneously than it is to overtake one system [93].

Blockchain also provided a solution more robust than the domain name system (DNS) for preventing DDoS attacks. One such method is simply to distribute the requests amongst the nodes in the network. When the network is overwhelmed by many requests, it distributes the requests amongst various nodes, rather than having a designated central node to handle requests. Thus, to overwhelm the network, the attacker would have to send enough requests to occupy all of the nodes simultaneously, without being flagged as a DDoS attack, which is much more complex than simply overwhelming a central server [102].

The application of Blockchain in cybersecurity can be divided into multiple ways. Blockchain applications for the following will be considered for this review – software-defined network specific solutions, internet-specific solutions, Internet of Things specific solutions, and cloud storage solutions.

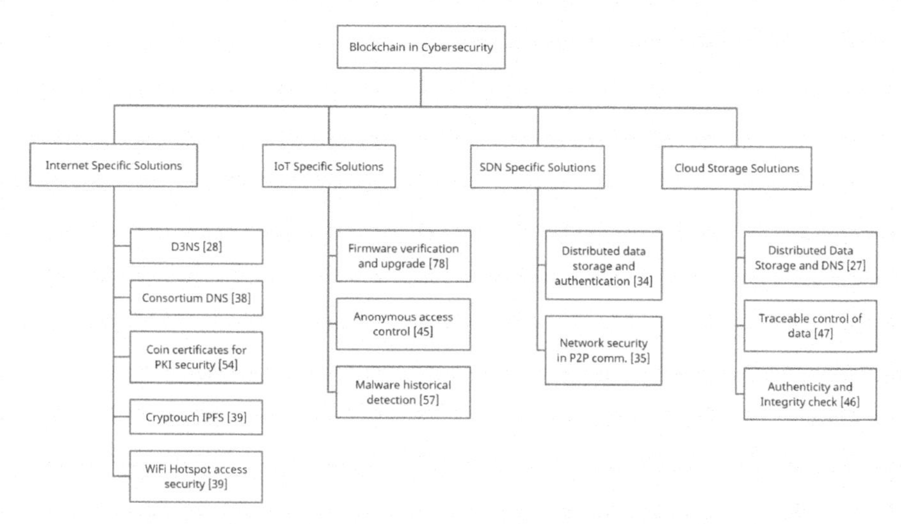

Figure 7.2 Applications of Blockchain in cybersecurity

7.4.2.1 Software-defined network (SDN) specific solutions

While Blockchain has been applied successfully in various network security applications, one of the most promising applications of Blockchain is to improve the security of Software-defined networks (SDNs). SDNs are a technique for building more robust and secure [12]. In SDN, the control plane, which decides how network packets should be forwarded, is logically centralized in a software entity called the SDN controller. This controller allows administrators to program and configure all the network elements from a single management point, removing the need to access every network device. This system was meant to prevent, detect, report, isolate, minimize, and mitigate the harmful effects of most network intrusions [19].

While SDNs do provide more security against some strengths, scalability, and ease in monitoring, they introduce new vulnerabilities that did not exist in traditional networks [22]. SDNs have seven main attack vectors, and the management and configuration of the centralized controller are considered the single point of failure [94].

As we have seen, Blockchains provide a method to secure decentralized systems. In applications with Blockchain-enabled SDNs, critical data are stored and authenticated in containers[34]. Both private and public Blockchains can be used for the P2P communication between the nodes in the network and SDN controllers, allowing the Blockchain to address the network security concerns presented by the SDN [35].

7.4.2.2 Internet-specific solutions

A large population of the world uses the internet to some extent every day. With its benefits, the internet also brings the concern of security. The development of Blockchain has led to a revolution in internet security. This subsection will discuss the security of the internet regarding the navigation and the use of the World Wide Web. In contrast, the two subsequent subsections will discuss more specific internet use, namely IoT and cloud.

The domain name system (DNS) is a centralized distribution database where IP addresses are mapped to domain names, allowing more accessible access to the internet. While the DNS is an essential part of the internet, it is cumbersome, inefficient, and is often the target of security attacks [95].

Benshoof *et al.* [28] proposed a distributed decentralized domain name system (D3NS) that utilizes Bitcoin Blockchain-based distributed hash tables and domain name ownership system. D3NS provides solutions to some identified current DNS problems such as DDoS, spoofing, and censorship. It provides a decentralized and authenticated record of domain name ownership and thereby eliminates the need for certificate authorities. Another notable Blockchain DNS solution was proposed by Wang *et al.* [38]; ConsortiumDNS is based on a consortium network containing miner nodes and query codes, with three-layer architecture and external storage to address and accommodate the Blockchain.

Another internet vulnerability was addressed by Qin *et al.* (2020). The public key infrastructure (PKI) allows public keys to be reliably linked to their owners, but the infrastructure is centralized, which raises the single-point failure concern. PKI is also vulnerable to attacks such as man-in-the-middle attacks. The authors proposed Coin, a Blockchain-inspired distributed certificate scheme, to tackle this issue. The system works by treating certificates as currencies and recorded on the Blockchain; miners verify the validity of certificates using a set of rules and allow an identity to bind multiple public-key certificates [54].

Saritekin *et al.* [55] proposed a model called Cryptouch that uses Blockchain and InterPlanetary File System (IPFS), a decentralized approach to the current HTTP protocol, to provide an intermediary, distributed, continuous, and secure communication environment. The IPFS, unlike the Blockchain, can handle large amounts of data and handle duplicates.

Niu *et al.* [39] proposed a model that uses the bitcoin Blockchain to provide an anonymous and accountable authentication scheme for Wi-Fi hotspot access. The model uses a P2P mixing protocol to exchange credentials recorded in the Blockchain for the user devices. Then, the hotspot providers could access the Blockchain to authenticate the users.

7.4.2.3 IoT-specific solutions

The Internet of Things (IoT) has allowed for significant advances in the automation of traditional manual systems. However, the basic principle of IoT – devices connected over the internet network – leaves these devices vulnerable to network attacks. IoT security has become a significant concern as IoT devices are integrated into various fields, including the medical field, consumer-based personal devices, smart homes, smart cities, and producer-based manufacturing units [16]. Traditional solutions generally implement security solutions via a server or a different device, as IoT devices themselves have typically low computing power. However, centralized devices create a single point of vulnerability and failure.

One of the most significant advantages of Blockchain-based IoT solutions is that they are decentralized and do not require an additional device for sharing computing. The Blockchain can be used with smart contracts (Section 4) to provide access control configured autonomously [45].

Traditionally, IoT devices that require firmware updates have faced many issues in the automation of the process due to security concerns. If illegitimate firmware is downloaded, the entire network of the device may be compromised. Blockchain solutions can also be used for firmware verification and updating IoT devices [78]. Each network device is considered a node and stores a copy of the Blockchain ledger. When a new firmware is introduced to a node, it is shared with neighboring nodes for peer verification through Blockchain consensus protocol. Once the firmware version and authenticity are verified, the nodes can install the firmware [56].

Another major issue faced by IoT devices is malware detection. IoT devices generally do not have enough computing power to employ all security solutions on themselves to identify other malicious nodes in the network. However, a Blockchain-based architecture could store historical connections in the ledger and sit between the application and transport layers [57]. Then, malicious connections are stored in the publicly available ledger, and new ones can be identified via majority voting by the rest of the nodes [58].

7.4.2.4 Cloud storage solutions

Cloud computing and cloud storage have revolutionized shared storage solutions; however, they still had a single point of failure in the ecosystem. Both private and public Blockchains can distribute the data and act as a naming system [27]. Further, the Blockchain allows for data exchange where a receipt can verify the data and is protected against unauthorized changes [46]. Additionally, it can use client-side encryption to enable data owners to have traceable control of the data [47].

Apart from this, the Blockchain can also be implemented to protect the ecosystem from DDoS attacks.

7.4.3 Smart Contracts

Like many mechanisms proposed before the Blockchain, smart contracts have been improved by introducing Blockchain. Szabo [2] proposed smart contracts to embed contractual clauses into the hardware and software of the involved system to make the break of contract expensive for the breacher by automatically enforcing contractual clauses as soon as a certain condition is satisfied.

7.4.3.1 Blockchain-based smart contracts

One way to implement smart contracts is on top of the Blockchain in executable computer programs. These programs may also have logical flows that preserve the connections between contractual clauses (e.g., conditional statements). Any execution of the contract is then recorded as an immutable transaction that is stored in the Blockchain. This can be a symbiotic relationship, as the smart contracts can, in turn, enforce contracts and access control [96]. The life cycle of Blockchain-based smart contracts can be divided into four phases – creation, deployment, execution, and completion.

Creation involves negotiation between the parties regarding the legally binding terms of the contract. These terms are then converted into executable software programs through design, implementation, and validation [30].

Deployment is performed once the smart contract has been validated sufficiently, and deployment is done on top of Blockchains to the platform. Once the Blockchain is deployed, all involved parties can access the contracts. All involved digital assets are locked by freezing their digital wallets – this ensures that the parties will be able to pay the penalty if the contractual clauses are broken. After deployment to the Blockchain, the contracts are immutable and cannot be changed – any modifications would require creating a new smart contract [40].

Execution of the smart contract occurs when any of the clauses are broken. For this, the contract is monitored and evaluated; when any of the multiple conditions are triggered, the corresponding statement is executed that causes a transaction; this is validated by the miners and then stored on the Blockchain [31].

Completion of the contract is broken either by fulfilling the conditions of the contract or by breaking the contract and triggering a transaction. Either way, the new states of the involved parties are evaluated, and the transactions that were a part of the smart contract are updated on the Blockchain. Then, the digital assets of all involved parties are unlocked, and the Blockchain is said to have completed its lifecycle.

7.4.3.2 Applications

Blockchain-based smart contracts provide a decentralized, authenticated, contractually binding platform that can be useful in many application areas.

7.4.3.2.1 Internet of Things

Blockchain has opened new avenues for IoT devices, and since smart contracts are automatically executed, they can take this development further. They can assist in IoT-based manufacturing or e-business transactions. For example, smart contracts can be programmed to update IoT-based devices without human intervention. Through the Blockchain, this system can be deployed across the whole network – the individual devices will obtain the firmware hashes from the smart contracts automatically and update them automatically [29].

Smart contracts also provide a failsafe for broken contracts and, thus, can be used to remove human interactions from transactions that would normally require a trusted third party. They can also be used to speed up traditional supply chains by automating payments and automatically charging penalties for breaches of the contract. Zhang *et al.* [26] proposed Distributed Autonomous Corporations that use smart contracts with Blockchain to safely automate transactions without the intervention of third parties such as governments, banks, or corporations.

7.4.3.2.2 Distributed system security

The distributed nature of the Blockchain has allowed for serious security advances and a shift in the way traditional networks are perceived. However, Blockchain technology can also be vulnerable to attacks. Smart contracts can provide an additional security mechanism over the Blockchain. Rodrigues *et al.* [41] proposed a method to mitigate DDoS attacks by automatically adding the IP addresses of the attackers into a smart contract that is then immediately available to other nodes in the Blockchain. The nodes can immediately enforce other security policies to mitigate the attack.

Further, smart contracts can be utilized to improve cloud security as well. Notably, they bypass the third-party broker that traditionally checks and matches user requests with service providers via smart contract-based service level agreements[77].

7.4.3.3 Finance

Smart contracts, like the Blockchain, can be utilized in various financial applications. They can be used to expedite settlement cycles in investment banking by automating the process [42]. Further, they can be used to implement mortgage or loan systems where the Blockchain automates payments, and any breach of the contract is reimbursed immediately to the loaner. Further, smart contracts can be used between companies and their users to form agreements on deliverables. For example, if the flight is delayed beyond the agreed time, the smart contract is triggered, and the customers are reimbursed.

7.4.3.4 Data Privacy and Reliability

Due to a shift toward machine learning and data science, many users need to be stored securely and protected from falsification. This is just one example; the methods for protecting and verifying user-generated data can also be used to protect other data sources such as intellectual property and records of transactions in a shared economy.

Depending on the application's requirements, smart contracts can be used to preserve the privacy of critical data while the Blockchain allows for a zero-knowledge approach to user privacy [43]. Smart contracts can also ensure that intellectual property, if stolen, can be traced to the leaker. Any shared data can be "watermarked" or appended to identify the transaction that occurred between the buyer and seller (such as their respective names or the unique ID of the payment). If the watermark appears on any platform outside the agreed usage domain, the smart contract will be triggered. Further, smart contracts can be used to verify that intellectual property has not been tampered with by embedding the data in the Blockchain for storage and checking newly received data with the embedded data for tampering [44].

7.4.4 Shortcomings of Blockchain Solutions in Cybersecurity

While Blockchain has a variety of applications, it also brings its drawbacks that must be considered when designing a Blockchain system. One of the significant limitations of Blockchain is the requirement to store a huge amount of data – as it stores all the transactions over time [59]. This reduces the distribution speed of the Blockchain over the network and also demands large storage space from the devices – which is often beyond the capacity of IoT or user devices. In addition, a restriction on block size and time interval reduces the transaction throughput of the network. Existing work redesigning the Blockchain and optimizing storage has not yet achieved a fully scalable solution [60].

Further, a distributed storage mechanism creates a broader attack surface in the Blockchain, giving attackers various alternatives to access the data that, although immutable, can be studied for sensitive information using techniques such as data mining [79].

There is also no consensus on the acceptable size of a block. While bigger blocks can store more data, they are difficult to store and distribute. On the other hand, smaller blocks are easy to manage and handle, and there is limited space to handle information.

The Blockchain, like any security mechanism, is itself vulnerable to attacks. Saad *et al.* [80] identified 22 different attacks related to the Blockchain structure, the peer-to-peer system, and the Blockchain application. Some of these attacks have been discussed here.

In addition, the consensus mechanism of the Blockchain is based on the assumption that the majority nodes in any network will be honest. However, suppose more than 50% of the minors can be controlled. In that case, the Blockchain process can be hijacked, and there is a possibility that even previously made transactions can be revised by providing a falsified history. Such attacks are commonly called majority attacks [18].

Even without control over a majority of the nodes, an attacker can carry out a successful attack such as selfish mining. Selfish mining exploits the tendency of the Blockchain to prefer longer chains. If an attacker were to append mined blocks to a private chain instead of broadcasting them until the private branch is longer than the public chain, their chain would replace the public chain. This would redirect mining awards to the attacker, to the detriment of miners from the original public chain [23]. Many selfish miners may come together to increase their profit regularly by consistently mining longer chains than the public chain.

Even though Blockchain is considered to mitigate DDoS attacks, Blockchain-based applications such as Bitcoin and Ethereum have repeatedly suffered from them [105]. Blockchain-based DDoS can be implemented in various ways, for example, the majority attack discussed above. Intentional malicious forking of the Blockchain can also lead to hard forks, which results similar to DDoS attacks. Further, due to the mining-based nature of the Blockchain, the number of transactions an application can process at a given time is limited. This can be exploited by introducing many transactions in a short period to overload and congest the network [93]. Further, the mempool can be flooded with unconfirmed transactions to increase mining [3].

7.5 Machine Learning Techniques

7.5.1 Introduction

Machine learning encompasses a broad range of algorithms and modeling tools that are taught to improve their performance through experience and data automatically. These algorithms can be used on specific training data to solve various problems in various disciplines. There are two fundamentally different types of models: generative models and discriminative models. Cybersecurity generally deals with the latter, which can further be classified into two tasks: classification, and regression. The classification works by assigning categorical or discrete class labels to the data. At the same time, a regression can be used to predict numerical or continuous values that represent a quantity dependent on the training data.

Machine learning algorithms can further be classified by the data used during their learning process. All models require some training data; however, unsupervised models learn on the training data alone while supervised models are provided with labels. Reinforcement falls between the two, where the model is provided with certain feedback that it can use in the training process [89].

The main idea behind the application of discriminative machine learning models to any field is that the models can find patterns in historical data applied to new data.

7.5.2 Applications in Cybersecurity

The current cyberspace generates, communicates, and stores incredible amounts of data. While traditional security solutions such as antivirus, firewalls, user authentication, access control, data encryption, cryptography systems, etc., are still incredibly important, they are not enough to fulfill the needs of the cyber industry [32].

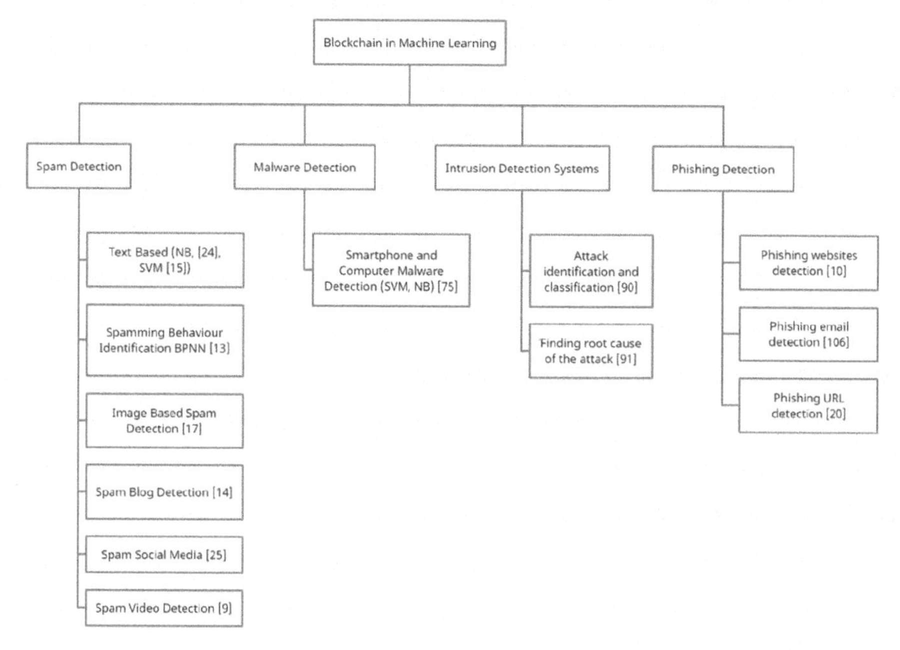

Figure 7.3 Applications of machine learning in cybersecurity

7.5.2.1 Intrusion detection systems

Intrusion detection systems (IDSs) are typically represented as *a device or software application that monitors a computer network or systems for malicious activity or policy violations* [21]. An IDS analyzes security data from several key points in a computer network or system.

There are four well-known IDS approaches, namely signature-based IDS, which can detect known attacks, anomaly-based IDS, which can detect new and unseen vulnerabilities, hybrid-approach IDS, which combine signature-based and anomaly-based approaches to reduce the false positive rate when detecting unknown attacks, and stateful protocol analysis approaches, which know and trace the protocol states [74].

Traditionally, security professionals employed file hashes, custom-written rules, or manually defined heuristics to make detections based on data obtained in the form of files, logs, network packets, or other relevant sources [6]. However, since both IDS and machine learning focus on the data, their technologies are often combined to create security solutions. The machine learning algorithms can learn or extract insight into security incident patterns from the training data for their detection and protection. This serves multiple purposes; automating the process makes it much faster than traditional approaches, and identifying unknown attacks significantly improves.

Classification techniques in IDS are generally used to detect whether or not an attack has happened or to classify attacks based on identified types based on data retrieved from the network or system. Various algorithms have successfully been used to create classification-based IDS, with support vector machines, decision trees, Naïve Bayes, Xero, OneR, and K-nearest neighbors being the most popular techniques [90].

Regression techniques, on the other hand, are used to predict continuous or numeric values dependent on the data, such as predicting the total number of attacks in a given time period, predicting network packet parameters, or even detecting the root cause of the attack [91].

7.5.2.2 Spam detection

Spam detections are usually applied on incoming emails, but other targets exist such as blogs, search engines, and even tweets. There are various bases used for creating spam detection techniques.

The first category is text-based classification, where the textual keywords in known and identified spam messages are used to train models to identify spam messages in the future. Bayesian classifiers that rely on a bag-of-words model are popularly used for spam detection [24]. Support vector machines (SVMs) are also popular for performing text-based classification [15].

Text-based classifiers had a huge drawback. Since they depended on keywords, they were easy to work around and grew outmoded with time. Another method focuses on "spamming behaviors" to characterize and detect spam messages. Back-propagation networks used for spam filtering have shown low misclassification rates with high accuracy [13].

Spammers began embedding spam messages in images to work around text-based spam detection. Computer vision and pattern recognition techniques have effectively been used to identify such images [17].

SVM is a popular technique used in various kinds of spam detection, including spam blogs [14], hacked or bot social media accounts [25], and spam videos on YouTube [9].

7.5.2.3 Malware detection

Detection of malicious code often analyzes code patterns and similarities. Unsupervised methods were initially used to detect malware based on access to the computer [7]. Various machine learning algorithms have been used for

Malware detection in both smartphones and computers, with SVM, decision trees, and Naïve Bayes being the most commonly used [75].

7.5.2.4 Phishing detection

Phishing is a form of social engineering where attackers masquerade as legitimate authorities and try to receive confidential or personal information from victims. SVM has been used to successfully classify phishing websites [10], detect phishing URLs[20], and detect phishing emails [106]. There is a relationship between the performance of SVM, decision trees, and Naïve Bayes, as they have also been used successfully for phishing detection, with variations in their performance based on the use case [53].

7.5.3 Shortcomings

While machine learning has produced remarkable results in various applications, there is still scope for improvement. Often, machine learning solutions involve a large amount of data that needs to be fed through the system. One of the biggest challenges with machine learning for cybersecurity is that the passage of time makes the data on which the model was trained increasingly redundant. Datasets from 1999 (NLK-KDD'99) are still used for research in 2021 [92]. Further, frequent changes in technology in cyberspace mean that models must be trained on new data to stay relevant; that is, pre-trained models will flag new packets as abnormal if they are not retrained. Finally, machine learning in cybersecurity often comes at the cost of context-awareness, which results in an uncertainty of the information gathered to discover attacks [33].

Even if vulnerabilities that could be secured with machine learning solutions are identified, research toward those solutions is impossible until a significant amount of attack data has been collected [100].

7.6 Integration of Machine Learning Blockchain Technology

Blockchain and ML are both up-and-coming technologies that have brought huge developments to cybersecurity. However, as seen above, both blockchain and ML have major scope for improvement. Interestingly, these two seemingly separate solutions have shown some scope for improving each other. This section discusses the same in two parts – how Blockchain can benefit machine learning and how machine learning can benefit Blockchain.

7.6.1 Blockchain to Improve Machine Learning

Machine learning solutions often deal with a large amount of data, either for training or sending through the algorithm for prediction. In these solutions, more significant amounts of data are preferred as they make the model more accurate and give more reliable predictions. However, collecting, organizing, auditing, safely storing, and accessing large amounts of data, especially updated regularly, require a huge overhead and are not a viable option for many systems. Blockchain can provide a method for storing and sharing data in a decentralized and tamper-proof way that the learning algorithms can access.

Doku and Rawat [82] proposed a permission-less Blockchain model that emphasizes quality over quantity and allows researchers to "buy" or "rent" data. The network is divided into interest groups, each of which has a unique dataset, and works on proof-of-interest rather than proof-of-work. These interest groups also make the model scalable.

The Ocean protocol is another decentralized Blockchain platform for AI data storage and transportation. The model uses three layers, and each of them works independently to provide access control, verification, and integrity to the system [97].

The ModelChain, proposed by Kuo and Ohno-Machado [65], is another decentralized solution that can preserve an ML framework on a private Blockchain. The transaction data is used for disseminating model and meta-information such as the flag, hash, accuracy, and error of the model. Each block has only one transaction with a unique timestamp, and the action flags the blocks the node has taken to the model – initialize, update, evaluate, or transfer. The incentive mechanism for this model is proof-of-information, and node priority is assigned like boosting [1]. The model provides security and robustness of privacy-preserving predictive models [65].

RoboChain, proposed by Ferrer *et al.* [69], is a learning framework used by multiple robot units. The robots connect to local hubs, interconnected in a network that validates the models or data. The robot plays the role of a node in this network and is responsible for sending transactions to the Blockchain that includes important information; along with this, the users can interact with the robots that ensure that the generated data do not compromise their privacy. This results in an efficient, secure, and decentralized mode for sharing data and models [69].

Androulaki *et al.* [66] provide a method for preserving the privacy of data sources used by machine learning models. Blockchain has a pseudonym

Table 7.1 Overview of the applications in which Blockchain was used to improve machine learning technologies

ID	Authors	Proposed model	Innovation	Security	Incentive	Scalability	Features	Future work
[65]	Kou and Ohno-Machado	ModelChain	Online ML framework on private Blockchain	Immutable audit trails, Mitigating Byzantine Generals and Sybil attacks	Proof-of-information	P2P modularity	Integration with existing public health systems Dissemination of partial models to nodes	Computational complexity of proof-of-information Majority attacks Scalability of proof-of-information Smart resource distribution Encryption of metadata Integration with VPNs
[68]	Wang	Unified analytic framework for ML on Blockchain	Implementation of ML as one model running in multiple threads on multiple machines	Basic Blockchain technology paradigms	PoW	Scalability in Blockchain nodes	Smart contracts to ensure fair play, evaluate models, and distribute rewards	Building incremental models Handling fast data streams
[69]	Ferrer *et al.*	RoboChain	Robot-unit-based learning network	Each node maintains data, privacy entropy calculated for shared answers	PoW	Future work	Feasible with low-cost robot units "Vetted" algorithms to reduce bias Deployable on any Blockchain, independent of specific implementation	Real-time learning Smart contracts to manage model proposal Modularity for handling multiple simultaneous interventions Compliance with different ethical standards
[70]	Weng *et al.*	DeepChain	Privacy concerns in federated learning	Individual threshold Paillier encryption by participants of obtained gradients Auditing of transactions and decryption shares Monetary penalty for failure to comply with timeout check	Monetary system established where rewards are earned by finishing a job first and becoming a leader. Money is deducted as usage fee of trained models or as penalties	Future work	Compatibility and liveliness in incentive mechanism Time-out checks Monetary penalties and rewards to increase fairness	Transfer learning Scalability
[82]	Doku and Rawat	Pledge	Permission-less Blockchain for "buying" and "renting" distributed data	Event Based Encryption, and Data Deletion	Proof-of-interest	Through interest groups	Information as a service	Efficient distribution of learning data amongst nodes Self-destructing mechanism for data Training on encrypted data or encryption of model
[83]	Shen *et al.*	secureSVM	Privacy preserving SVM training scheme over Blockchain-based encrypted IoT data	Local data encryption via Paillier	PoW	Achieved via modularity in data providers	IoT data analyst access inclusion Consensus mechanism to secure against tampering of data	Generalization to other models

(*Continued*)

Table 7.1 Continued.

ID	Authors	Proposed model	Innovation	Security	Incentive	Scalability	Features	Future work
[84]	Bore *et al.*	Decentralized trust between machine learning parties	Establishing distributed trust in the computations performed over multi-agent systems, using validation techniques and Blockchain logging	monitoring of outputs to rank and identify faulty workers, inbuilt cryptography	Historical reward spaces	Novel compression schema to improve scalability	Tracking, monitoring, validation, and verification of ML tasks and results Input/output cloud-enabled APIs User management Human management of results	Mitigation of attacks on the Blockchain Protection against bias
[85]	Lyu *et al.*	Fair and Privacy Preserving Deep Learning (FPPDL)	Incorporating fairness in federated learning	Three-layer onion-style encryption	None, parties expected to be law abiding and work fairly to get better models	Multi-fold optimization on the FL model	Each participant receives individual FL model reflecting their contributions Detection and reduced impact for low-contribution parties	Improved security during gradient sharing stage Quantification of fairness in non-regulated settings Mitigation of attacks on the Blockchain Making model deployable to real-world applications
[97]	Patnam and Gupta	Ocean protocol	AI data storage and transportation	Secret store functionality, digital signature algorithms	Cryptographic proof-of-service and network rewards	Future work	Right to privacy via service execution agreements	Integration of hardware-based trust mechanisms Security of secret store node Cryptographic techniques resistant to quantum computing

system that allows nodes to hide their identity in situations such as healthcare that require high privacy. Shen [83] proposed secureSVM, which encrypts the data locally before recorded on a distributed ledger. SecureSVM also removes the need for a third party by employing secure building blocks.

In the work of Lyu *et al.* [85], Blockchain was utilized to create a more robust and secure network for deep learning using two algorithms, initial benchmarking, and privacy-preserving collaborative. In the former, the various parties or nodes train a local model, a differentially private generative adversarial network (DPGAN). The generated samples are then published in the network for other parties to label based on local creditability and transaction points. The framework also utilizes a three-layer encryption scheme that preserves the system's accuracy, privacy, and robustness.

Another privacy technique is where the node devices store the data in their InterPlanetary File System. The ML algorithm stores the personalized parameters of each node on the Blockchain, with the hash of the data generated by the nodes. This system allows for users' customization without risking the privacy or security of users [67].

Blockchain can also be used to decentralize machine learning models traditionally built with centralized technologies in mind, where a single entity has access to resources for storing large datasets, creating sophisticated models, and using the results from the model to solve the issue at hand. However, this approach is limited by the huge amount of data to process, the single point of failure, and the fact that they are not scalable.

This decentralized mechanism can expand the application of machine learning in the context of the real world. Using a decentralized framework based on Blockchain, the authors achieved security and integrity without compromising the collaborative predictive analysis and intelligent decision-making among multiple UAVs [103].

The fundamental nature of Blockchain is to provide a decentralized system while taking security and privacy into consideration. An ML model decentralized based on Blockchain can provide storage for batches of verified data for training, protects against tampering, and allows the data to be available at multiple locations securely by demanding proof from the operators. The Blockchain can also provide reliable and secure methods to share model parameters and metadata, such as which batch of data was used to train the model [84].

Blockchain can also be used to address the privacy concerns in federated learning, which supports uploading and sharing the model's training and knowledge, as shown by DeepChain [70]. It makes use of workers that are incentivized to process transactions and update local training and parameters, along with aggregating local intermediate gradients from untrusting parties through creating transactions. It also uses cryptographic techniques to ensure confidentiality and auditability. It also employs mechanisms such as collaborative training, time-out checking, and monetary penalty to push for fair behaviors amongst participants.

The implementation of modern learning algorithms often depends on cloud databases to store the huge amount of data used for training and testing; however, such databases often lack a convenient integrity check for the data. Vainshtein and Gudes [104] propose an interaction between a PoW-based Blockchain and the cloud platform to utilize a distributed hash table and lightweight software units that monitor any changes done to the cloud database storage nodes. These software units publish any database operations to the Blockchain audit. The cloud provider is given access to the metadata to check for corruption of transactions and handle these incidents when they are detected.

Finally, Blockchain can also imbibe more trust in the decision-making process for ML models. As the algorithms become more advanced, the process involved in their decision-making becomes more complex and difficult for humans to understand. Blockchains can be used to record both the decision-making process of the algorithm and the results, which are immutable and transparent so that anyone can view them and verify or audit the decision-making process [50].

Wang [68] proposed a framework that provides trust by utilizing five parts: the core ML algorithm that will run on the data; the server and streaming layer that host computing environments for the algorithms; a smart contracts layer that hosts the ML activities; and an append-only Blockchain that stores the results of the framework. Hence, a single machine can implement multiple threads on different machines via the Blockchain and later access the results from the same.

Essentially, Blockchain can solve an age-old problem in machine learning that can also be applied to big data and data science. Its ability to secure transactions and provide a decentralized framework warrants a large amount of attention and indicates rapid development in the future, as will be discussed in the future work section.

7.6.2 Machine Learning to Improve Blockchain Solutions

As discussed, Blockchain solutions are still vulnerable to various attacks, and ML provides a possible solution to mitigate some of these. ML algorithms can be used to detect attacks on the Blockchain so that preventative measures can be taken. For example, machine learning can be applied to the Blockchains that handle and share sensitive data [64]. Further, the results can be shared across the nodes of the network. Once a consensus is reached, the data is added to a new block within the network after encrypting the data, and the keys are only distributed to select parties [49].

Anomaly detection systems for Blockchains can also be implemented because attacks that occur may also happen on different network nodes. Unlike other nodes, this model uses information about orphans and forks to make decisions. When a node identifies that it is attacked, it shares the information in the network, and the other nodes use machine learning algorithms to detect similar attacks [63].

Machine learning can also be implemented to ensure the secure sharing of data across the network between end-users and a virtual node. Tsolakis *et al.* [62] proposed a system where fog-enabled intelligent devices [81] are

Table 7.2 Overview of applications in which machine learning techniques have been used to improve security in Blockchain models, where the last two rows focus primarily on smart contracts

ID	Authors	Proposed model	Innovation	Features	Future work
[48]	Yin and Vatrapu	NA	Entity recognition to identify criminals based on data regarding transaction	Clustering based on co-spend, intelligence, and behavior Identification of 12 different categories of entities and transaction types	Improve accuracy of models Consider class imbalance in data Automate data cleaning process
[52]	Wasim *et al.*	LaaS: law as a service	Probability-based factor model imple-mented over Blockchain to identify breaches in online contracts	Automatic issue of court injunction Built upon factor loading and stochastic modeling Protection against non-repudiation	Capability with Memcached Resistant to failure if *t*-test fails
[61]	Chawathe	Self-organizing maps for Blockchain monitoring	Modular security tool to be used in conjunction with other tools	Identify abnormal behavior Implementation of data manager and interactive visualizer	Generalization for Blockchains other than Bitcoin Improved efficiency
[62]	Tsolakis *et al.*	OpenADR2.0: Secure and trusted demand	Interoperable and secure energy transactions through the	Use of DELTA security framework Mitigating	Generalizations to other Blockchain models

(*Continued*)

Table 7.2 Continued.

ID	Authors	Proposed model	Innovation	Features	Future work
	response (DR) system	based on Blockchain technologies	combination of an open DR system	risk of fluctuating energy prices Autonomous P2P contracting Access of smart contracts though dedicated decentralized applications	Optimization of security-efficiency tradeoff
[63]	Signorini *et al.*	BAD: Blockchain anomaly detection	Historical fork data to blacklist attack features	Collect local attack logs as hashed malicious transactions A threat database for reference against attacks	Deployment on real networks Testing based on attacks other than Eclipse
[71]	Chen	Takagi–Sugeno fuzzy cognitive maps ANN	Traceability chain algorithm with reduced dimension-ality	Both forwards and backwards auditing Elimination of irrelevant data Quantification of impact of consensus	Improvement of deep learning integrated hash-cash algorithm Testing with other fuzzy techniques Incorporation of domain knowledge
[107]	Chen *et al.*	Proof-of-artificial intelligence	AI-based super node	Less resource consumption than three	Development into a complete

Table 7.2 Continued.

ID	Authors	Proposed model	Innovation	Features	Future work
			selection technique using CNN to calculate the average transaction number and select those above the threshold	major proof algorithms Node capability mechanism to shorten cycle of reaching consensus Exploits nearly complimentary information of each node Dynamic threshold used to ensure safety	consensus protocol Integration with various existing Blockchain networks
[108]	Luong *et al.*	Optimal auction in mobile Blockchains	Edge computing resource management based on optimal auction using deep neural networks	Maximization of revenue for edge computing service providers Miner valuations used as input data Optimization of losses in the network	Generalization with multiple resource units ANN construction without characterization results
[110]	Demertzis *et al.*	Deep learning smart contracts	Anomaly detection in Blockchain security architecture using deep autoencoder	Bilateral traffic control Reliable, low-demand anomaly detection system Real-time critical	Further optimization of deep agreement autoencoder parameters Improvement in classification

(*Continued*)

Table 7.2 Continued.

ID	Authors	Proposed model	Innovation	Features	Future work
				defense mechanism Calculation of optimal dataset threshold to improve classification results	of states of IIoT devices Automation of re-training and improvement techniques Addition of a cross-sectional anomaly analysis system Incorporation with Big Data techniques

implemented on the user side, and smart contracts acted on at the producer side to ensure safe transactions.

Wasim *et al.* [52] proposed a probability-based factor model (PFM) using unsupervised machine learning that performs factor analysis and stochastic modeling. The model is implemented over the Blockchain to perform two-factor authentication, identify data breaches, and predict the probability of such breaches occurring in the future. In addition, machine learning algorithms can also be trained to identify breaches in the Blockchain. Some Blockchains may warrant specific algorithms trained on their environment to work efficiently: In the work of Yin and Vatrapu [48], a model was trained specifically on data including cyber-criminal activities in the bitcoin architecture, and the prototype was capable of producing an analysis and also filtering the list of suspicious addresses by a number of categories based on previous investigations. The prototype can, therefore, both identify the crime committed by an address and flag addresses as likely to commit fraud in the future.

Further, ML can be used for entity recognition in the Blockchain. Classification models can be used to classify the activity on the Blockchain as normal or malicious, with malicious activities may be broad or specified [48].

The algorithms can then be implemented in the Blockchain layer based on the specific application (for example, DDoS attacks by majority mining would need to identify malicious nodes before the chain is updated). Machine learning algorithms can also be used to identify specific use cases such as nodes that are related to ransomware.

One of the major issues arising from Blockchain anonymity is the difficulty in tracing fraudulent transactions. Artificial neural networks, along with fuzzy cognitive maps, were utilized by Chen in [71] to act as a traceability chain algorithm. With the use of the inference mechanism, the algorithm can discard frivolous data and run faster than consensus algorithms.

The major application for machine learning in Blockchain-based applications is the mitigation of attacks at the various levels of the Blockchain. It can be used both to detect and prevent attacks, based on what the algorithms are trained to identify. Future work is required in improving security applications based on Blockchain.

Blockchains can also be implemented with self-organizing maps (SOMs) that act as a monitoring system and remove the need for external control. ML algorithms can identify patterns similar to or partially recreations of previously identified events. They can also communicate with neighboring nodes to detect events within the network [61].

Another common issue faced by Blockchain networks is energy efficiency issues. To address these issues using convolutional neural networks, Chen *et al.* [107] proposed a model called proof-of-artificial intelligence (PoAI) that consists of two types of nodes: random nodes and supernodes that are more computationally powerful. The supernodes can reach consensus faster, but random nodes exist to guarantee fairness; both are assigned by a CNN framework that extracts information of each node, calculates the predicted average transaction of each node, and assigns nodes as supernodes or random nodes based on threshold values. The framework considerably shortens the time for reaching consensus.

The problem of resource allocation and energy consumption is magnified in mobile Blockchains due to the weak computational power of nodes. Luong *et al.* [108] propose an auction-based edge-resource allocation mechanism that allows an edge-computing service provider to offer offloading mining tasks from mobile devices. The miners' bid for the service and a multi-layer neural network is constructed and updated autonomously to converge to a solution for energy and resource efficiency quickly.

7.6.2.1 Machine learning applications in smart contracts

As discussed in Section 4, smart contracts are often employed in Blockchain applications for an added layer of security and protection against non-repudiation. Although smart contracts have shown much promise, the implementation of fully automated smart contracts in real-world applications is complicated due to a lack of legal viability, constraints born from code-immutability, and, as mentioned before, the complexity of implementation. Smart contracts on their own do not forgive to the ambiguity, flexibility, and uniqueness that many contracts require. Here, machine learning can help improve smart contracts: a promising solution is offered by natural language processing, a sub-set of machine learning, which can perform tasks such as named entity recognition, semantic parsing, and word sense disambiguation, which can automate the requirement generation process of smart contracts.

ML can also be applied in smart contracts to analyze the history of transactions and map a pattern of frequent bidders to optimize the functionality. For example, McMillan [109] used neural networks to analyze data from a massive network of IoT devices and map the data into weighted relationships that could be studied to understand the energy infrastructure and requirements of the network. This has the potential for automating auctioning in energy markets, which can be applied to optimizing Blockchain energy resources. Demertzis *et al.* [110] proposed an infrastructure to handle anomaly detection in industry 4.0 meant for Industrial IoT via deep autoencoder neural networks (DANNs) that are trained in a Blockchain framework that uses smart contracts for bilateral traffic control agreements. The neural network acts more as an active structural element than a supporting framework. The proposed system comprises three layers, authorization, syndication, and overlay, and a fully decentralized authorization system that operates globally. The authorization layer provides various access levels by expressing security policies, using resources, namespaces, entities, and delegations of trust. An entity is an independent unit that can grant, reset, or delete access rights. The syndication layer acts concerning the authorization layer and provides the publish/subscribe functions to various resources. The overlay layer forms the network between the IoT devices, including the communication network and the Blockchain network, over the existing physical layer. When the smart contract is triggered by a rule in this system, such as one device requesting to send information to another, the smart contract routes the contact to a cloud service where the features are extracted and presented to the trained DANN. The model predicts whether the action or traffic is normal or abnormal based on a reconstruction error threshold. If it is abnormal, the transaction is hidden from the network, and an alert is sent to the security operations center; otherwise, the traffic is sent forward as normal.

7.7 Future Work

While machine learning and Blockchain applications have been significantly considered in the cybersecurity context in recent years, the field is relatively new. Both involved technologies are also seeing a high rate of development today. Future research is also required toward making combined Blockchain–ML solutions more robust, scalable, and deployable to even current applications. This section is dedicated to identifying future research directions for integrating machine learning techniques and Blockchain-based networks to improve cybersecurity.

As discussed in the previous sections, one of the most significant issues faced in a cybersecurity context is that the real-world scenario changes daily. The solutions implemented to protect against one attack often expose the network to other vulnerabilities; in this scenario, the Blockchain is vulnerable to attacks, especially in the application layer. While machine learning can and has been trained to identify breaches in Blockchain security, to this date, it is not possible to train machine learning algorithms on large amounts of high-dimensionality data. Furthermore, either the data fed to the training model needs to be manually labeled or the output from the model needs to be manually labeled: unsupervised learning without an understanding of the results cannot be realistically applied to security applications. Here lies the major unresolved issue of machine learning in cybersecurity applications in general. As long as attackers can find new attacks, pre-trained machine learning algorithms will become less effective, and constant retraining of algorithms is unfeasible due to resource and data constraints. Further research is required to make machine learning algorithms more scalable, not just in a spatial sense but also in a temporal aspect.

As mentioned in previous sections, the efficiency of the consensus protocol still requires refinement, and research has been conducted toward constructing a suitable consensus mechanism for Blockchain applications. However, the full utilization of machine learning in consensus mechanisms has not yet been discussed. In the model proposed by Bravo-Marquez [111], the consensus mechanism does employ machine learning for a consensus mechanism in which resources are used to solve practical machine learning tasks rather than merely solving hash-based puzzles. However, there is a need for a collision prevention mechanism in this setup. Other research has been done toward learning-based consensus tasks, such as Chenli [51]; however, the proposed model carries a large latency of block generation and is vulnerable to attacks by malicious requesters. Other issues faced during research into a learning-based consensus mechanism found it difficult to find a tradeoff between the complexity of the training and the effort taken to reach a consensus, as a higher complexity of training would increase latency and require more resources. Still, a lower complexity would make it easier to fake a majority consensus.

Another major issue faced during Blockchain development is scalability. In a traditional network, as the size of the Blockchain increases with time, each node is expected to store the entire Blockchain: in fact, the longest chain is often preferred. Most traditional networks cannot, thus, handle a high number of transactions quickly without incurring high costs, which is a problem as Blockchains are gaining popularity. Some techniques include splitting the Blockchain into more manageable sizes, such as the lightning protocol and Blockchain sharding [112]. Sharding partitions the entire network into shards such that each contains its state and transaction history. The nodes that manage a particular shard maintain information only on their respective shard: they do not need to load the Blockchain, effectively splitting the chain into parallel sub-chains. ML solutions could work perfectly in this scenario for the sophistication of sharding techniques. Further, they could be used to intertwine sharding and pruning solutions. The algorithms can identify which transactions can be dropped from the Blockchain because they are sufficiently in the past without affecting the system's security.

Further, many critical real-world systems that could benefit from the solutions mentioned in this chapter have infrastructures that are not compatible with recent developments. The Blockchain requires the presence of several decentralized nodes in a large enough quantity to support the consensus mechanism. Further, apart from decentralized storage sufficient to host the Blockchain, the infrastructure should house the model and data for the learning. Also, while the Blockchain does not require a network administrator, the use case may require a central administrator or quality control position, which must be accommodated for. Finally, some networks may require a hardware upgrade to host the communication protocols required by the Blockchain. These changes can be costly, but even if the benefits outweigh the costs, they may require that the system be offline for a period of time, which is not possible for some critical applications. Research toward creating optimized products for use in Blockchain networks is underway, including virtual machine infrastructures; developments in this technology should take the business aspect of implementing solutions into account.

As mentioned before, smart contracts are automated agreements that are not yet legally binding but are expected to secure the Blockchain against issues such as non-repudiation. They are meant to trigger automatically and deduct a "fine" from the offender if a rule specified in the smart contract is broken. However, while some repeating contracts can be automated to trigger a specific action, many smart contracts need to be designed from scratch to accommodate changing requirements or expectations; i.e., they need to be negotiated on and coded and then tested for functionality. This is a convoluted and time-consuming process. Regnath and Steinhorst [73] proposed the creation of natural-language-based smart contracts that are more easily understandable by human readers; SmartCoNat, while utilizing primarily human-readable language, also included strict rules such as a clear code structure and a limit to aliasing to avoid the ambiguity that is usually present in natural languages. Despite making the contract more human-readable, the language neither simplifies the creation of the rule nor automates it. While research has already been done into automating the coding of smart contracts, there has been some indication that machine learning techniques such as natural language processing may accomplish much more. Smart contracts have the potential of being fully automated, including negotiations using natural language processing. These systems would either parse the entire conversation between the users and service providers or carry out the negotiation process as a whole on their own, with the best interests of their representative party. Of course, such a system is presently purely speculative and would require further development in natural language processing to be feasible.

7.8 Conclusion

In conclusion, Blockchain and machine learning have individually resulted in remarkable developments in security applications. However, when used individually, each is vulnerable to attacks and does not manage resources as efficiently as possible. When combined, the two technologies, can work to eliminate, or at the least reduce, the shortcomings of the other. Future research is required toward making combined Blockchain–ML solutions better. Particularly, (1) the scalability of solutions, (2) generalizations to multiple use cases, and (3) method of integration with existing real-time systems are the most pressing avenues for future research.

References

[1] Freund, Y., and Schapire, R.E. (1996). Experiments with a new boosting algorithm. In *Proceedings of ICML*. 96, 148-156.

[2] Szabo, N. (1997). Formalizing and securing relationships on public networks. *First Monday*, 2, 9. DOI: 10.5210/fm.v2i9.548.

[3] Blaze, M., Feigenbaum, J., Ioannidis, J., and Keromytis, A.D. (1999). The role of trust management in distributed systems security. *Secure Internet Programming*, 1, 185-210. DOI: 10.1007/3-540-48749-2_8.

[4] Back, A. (2002). Hashcash—A denial of service counter-measure.

[5] CNN.com - MyDoom worm spreads as attack countdown begins, Jan. 29, 2004. Available at: https://edition.cnn.com/2004/TECH/internet/01/29/mydoom.future.reut/ [accessed Aug 13, 2021].

[6] Mukkamala, S., Sung, A., and Abraham, A. (2005). Cybersecurity challenges: Designing efficient intrusion detection systems and antivirus tools. Vemuri, V. Rao, Enhancing Computer Security with Smart Technology. 125-163.

[7] Stolfo, S.J., Hershkop, S., Bui, L.H., Ferster, R., and Wang, K. (2005). Anomaly detection in computer security and an application to file system accesses. In *Proceedings of ISMIS*. 5, 14-28. DOI: 10.1007/11425274_2.

[8] Chandrasekaran, M., Narayanan, K., and Upadhyaya, S. (2006). Phishing email detection based on structural properties. In *Proceedings of 9th Annual NYS Cyber Security Conference*, Albany. 2-8.

[9] Benevenuto, F., Rodrigues, T., Almeida, V., Almeida, J., Zhang, C., and Ross, K. (2008). Identifying video spammers in online social networks. In *Proceedings of the 4th International Workshop on Adversarial Information Retrieval on the Web*. 4, 45-52.

[10] Bose, A., Hu, X., Shin, K.G., and Park, T. (2008). Behavioral detection of malware on mobile handsets. In *Proceedings of the 6th International Conference on Mobile Systems, Applications, and Services*, Breckenridge, CO, USA. 6, 225-238. DOI: 10.1145/1378600.1378626.

[11] Nakamoto, S. (2008). A peer-to-peer electronic cash system. Available at: https://bitcoin. org/bitcoin. [accessed July 05, 2021].

[12] McKeown, N. (2009). Keynote talk: Software-defined networking. In *Proceedings of IEEE INFOCOM*, Rio de Jenerio, Brazil. 9.

[13] Wu, C.H. (2009). Behavior-based spam detection using a hybrid method of rule-based techniques and neural networks. *Expert Systems With Applications*, 36(3), 4321-4330. DOI: 10.1016/j.eswa.2008.03.002.

[14] Abu-Nimeh, S., and Chen, T. (2010). Proliferation and detection of blog spam. *IEEE Security & Privacy*, 8(5), 42-47. DOI: 10.1109/MSP.2010.113.

[15] Amayri O., and Bouguila, N. (2010). A study of spam filtering using support vector machines. *Artificial Intelligence Review*, 34(1), 73–108. DOI: 10.1007/s10462-010-9166-x.

[16] Atzori, L., Iera, A., and Morabito, G. (2010). The internet of things: A survey. *Computer Networks*, 54(15), 2787-2805. DOI: 10.1016/j.comnet.2010.05.010.

[17] Biggio, B., Fumera, G., Pillai, I., and Roli, F. (2011). A survey and experimental evaluation of image spam filtering techniques. *Pattern Recognition Letters*, 32(10), 1436–1446. DOI: 10.1016/j.patrec.2011.03.022.

[18] Barber, S., Boyen, X., Shi, E., and Uzun, E. (2012). Bitter to better—how to make bitcoin a better currency. In *Proceedings of International Conference on Financial Cryptography and Data Security*, Eds Keromytis, A.D., Springer, Heidelberg, Berlin. 7397, 399-414. DOI: 10.1007/978-3-642-32946-3_29.

[19] Tank, G.P., Dixit, A., Vellanki, A., and Annapurna, D. (2012). Software-defined networking: The new norm for networks. In *Proceedings of Tank2021SoftwareDefinedNN*. 2(2-6), 11.

[20] Jiuxin, C., Dan, D., and Tianfeng, M.B.W. (2013). Phishing detection method based on URL features. *Journal of Southeast University (English Edition)*, 29(02), 134-138. DOI: 10.3969/j.issn.1003-7985.2013.02.005.

[21] Johnson, L. (2013). *Computer Incident Response and Forensics Team Management: Conducting a Successful Incident Response*, Eds Kessler, M. Newnes. DOI: 10.1016/C2012-0-01092-7.

[22] Kreutz, D., Ramos, F.M., and Verissimo, P. (2013). Towards secure and dependable software-defined networks. In *Proceedings of the 2nd ACM SIGCOMM Workshop on Hot Topics in Software-Defined Networking*. 55-60. DOI: 10.1145/2491185.2491199.

[23] Eyal, I., and Sirer, E.G. (2014). Majority is not enough: Bitcoin mining is vulnerable. In *Proceedings of International Conference on Financial Cryptography and Data Security*. 436-454. DOI: 10.1007/978-3-662-45472-5_28.

[24] Wang, Z.J., Liu, Y., and Wang, Z.J. (2014). Email filtration and classification based on variable weights of the Bayesian

algorithm. In *Applied Mechanics and Materials*, 503, 2111–2114. DOI: 10.4028/www.scientific.net/AMM.513-517.2111.

[25] Zangerle, E., and Specht, G. (2014). "Sorry, I was hacked" A classification of compromised twitter accounts. In *Proceedings of the 29th Annual ACM Symposium on Applied Computing*. ACM. 29, 587-593. DOI: 10.1145/2554850.2554894.

[26] Zhang, Y., and Wen, J. (2015). An IoT electric business model based on the protocol of bitcoin. In *Proceedings of the 2015 18th International Conference on Intelligence in Next-Generation Networks*. IEEE. 184-191. DOI: 10.1109/ICIN.2015.7073830.

[27] Ali, M., Nelson, J., Shea, R., and Freedman, M.J. (2016). Blockstack: A global naming and storage system secured by Blockchains. *In Proceedings of 2016 {USENIX} Annual Technical Conference*, Denver, CO, USA. 181-194.

[28] Benshoof, B., Rosen, A., Bourgeois, A.G., and Harrison, R.W. (2016). Distributed decentralized domain name service. In *Proceedings of the 2016 IEEE International Parallel and Distributed Processing Symposium Workshops (IPDPSW)*. IEEE. 1279-1287. DOI: 10.1109/IPDPSW.2016.109.

[29] Christidis, K., and Devetsikiotis, M. (2016). Blockchains and smart contracts for the Internet of Things. *IEEE Access*, 4, 2292-2303. DOI: 10.1109/ACCESS.2016.2566339.

[30] Idelberger, F., Governatori, G., Riveret, R., and Sartor, G. (2016). Evaluation of logic-based smart contracts for Blockchain systems. In *Proceedings of the International Symposium on Rules and Rule Markup Languages for the Semantic Web.* Eds Alferes, J., Bertossi, L., Governatori, G., Fodor, P., Roman, D., Springer, Cham. 9718, 167-183.

[31] Koulu, R. (2016). Blockchains and online dispute resolution: smart contracts as an alternative to enforcement. *SCRIPTed*, 13, 40.

[32] Anwar, S., Mohamad Zain, J., Zolkipli, M.F., Inayat, Z., Khan, S., Anthony, B., and Chang, V. (2017). From intrusion detection to an intrusion response system: Fundamentals, requirements, and future directions. *Algorithms*, 10, 39. DOI: 10.3390/a10020039.

[33] Aleroud, A., and Karabatis, G. (2017). Contextual information fusion for intrusion detection: A survey and taxonomy. *Knowledge and Information Systems*, 52(3), 563-619. DOI: 10.1007/s10115-017-1027-3.

[34] Basnet, S.R., and Shakya, S. (2017). BSS: Blockchain security over software-defined network. In *Proceedings of the 2017 International*

Conference on Computing, Communication and Automation (ICCCA). IEEE. 720-725. DOI: 10.1109/CCAA.2017.8229910.

[35] Bozic, N., Pujolle, G., and Secci, S. (2017). Securing virtual machine orchestration with Blockchains. In *Proceedings of the 2017 1st Cyber Security in Networking Conference).* 1, 1-8. DOI: 10.1109/CSNET.2017.8242003.

[36] Schatz, D., Bashroush, R., and Wall, J. (2017). Towards a more representative definition of cybersecurity. *Journal of Digital Forensics, Security and Law*, 12(2), 53-74. DOI: 10.15394/jdfsl.2017.1476.

[37] Kshetri, N., and Voas, J. (2017). Do crypto-currencies fuel ransomware? *IT Professional*, 19(5), 11-15. DOI: 10.1109/MITP.2017.3680961.

[38] Wang, X., Li, K., Li, H., Li, Y., and Liang, Z. (2017). ConsortiumDNS: A distributed domain name service based on consortium chain. In *Proceedings of the 2017 IEEE 19th International Conference on High Performance Computing and Communications; IEEE 15th International Conference on Smart City; IEEE 3rd International Conference on Data Science and Systems (HPCC/SmartCity/DSS).* 617-620. DOI: 10.1109/HPCC-SmartCity-DSS.2017.83.

[39] Niu, Y., Wei, L., Zhang, C., Liu, J., and Fang, Y. (2017). An anonymous and accountable authentication scheme for Wi-Fi hotspot access with the Bitcoin Blockchain. In *Proceedings of the 2017 IEEE/CIC International Conference on Communications in China (ICCC).* IEEE. 1-6. DOI: 10.1109/ICCChina.2017.8330337.

[40] Sillaber, C., and Waltl, B. (2017). Life cycle of smart contracts in Blockchain ecosystems. *Datenschutz und Datensicherheit-DuD*, 41(8), 497-500. DOI: 10.1007/s11623-017-0819-7.

[41] Rodrigues, B., Bocek, T., Lareida, A., Hausheer, D., Rafati, S., and Stiller, B. (2017). A Blockchain-based architecture for collaborative DDoS mitigation with smart contracts. In *Proceedings of the IFIP International Conference on Autonomous Infrastructure, Management and Security* Eds Tuncer, D., Koch, R., Badonnel, R., Stiller, B., Springer, Cham. 16-29. DOI: 10.1007/978-3-319-60774-0_2.

[42] Panisi, F. (2017). Blockchain and smart contracts: FinTech innovations to reduce the costs of trust. DOI: 10.2139/ssrn.3066543.

[43] Xu, L., Shah, N., Chen, L., Diallo, N., Gao, Z., Lu, Y., and Shi, W. (2017). Enabling the sharing economy: Privacy respecting contract based on public Blockchain. In *Proceedings of the ACM Workshop*

on Blockchain, Cryptocurrencies and Contracts. ACM. 15–21. DOI: 10.1145/3055518.3055527

[44] Liang, X., Shetty, S., Tosh, D., Kamhoua, C., Kwiat, K., and Njilla, L. (2017). Provchain: A Blockchain-based data provenance architecture in a cloud environment with enhanced privacy and availability. In *Proceedings of the 2017 17th IEEE/ACM International Symposium on Cluster, Cloud and Grid Computing (CCGRID)*. IEEE. 468-477. DOI: 10.1109/CCGRID.2017.8.

[45] Pinno, O.J.A., Gregio, A.R.A., and De Bona, L.C. (2017). Control chain: Blockchain as a central enabler for access control authorizations in the IoT. In *Proceedings of the GLOBECOM 2017-2017 IEEE Global Communications Conference*. IEEE. 1-6. DOI: 10.1109/GLOCOM.2017.8254521.

[46] Yue, L., Junqin, H., Shengzhi, Q., and Ruijin, W. (2017). Big data model of security sharing based on Blockchain. In *Proceedings of the 2017 3rd International Conference on Big Data Computing and Communications (BIGCOM)*. IEEE. 117-121. DOI: 10.1109/BIGCOM.2017.31.

[47] Cai, C., Yuan, X., and Wang, C. (2017). Hardening distributed and encrypted keyword search via Blockchain. In *Proceedings of the 2017 IEEE Symposium on Privacy-Aware Computing (PAC)*. IEEE. 119-128. DOI: 10.1109/PAC.2017.36.

[48] Yin, H.S., and Vatrapu, R. (2017). A first estimation of the proportion of cybercriminal entities in the bitcoin ecosystem using supervised machine learning. In *Proceedings of the 2017 IEEE International Conference on Big Data (Big Data)*. IEEE. 3690-3699. DOI: 10.1109/BigData.2017.8258365.

[49] Alexopoulos, N., Vasilomanolakis, E., Ivánkó, N.R., and Mühlhäuser, M. (2017). Towards Blockchain-based collaborative intrusion detection systems. In *Proceedings of the International Conference on Critical Information Infrastructures Security*. Eds D'Agostino, G., Scala, A., Springer, Cham. 107-118. DOI: 10.1007/978-3-319-99843-5_10.

[50] Alexopoulos, N., Daubert, J., Mühlhäuser, M., and Habib, S.M. (2017). Beyond the hype: On using Blockchains in trust management for authentication. In *Proceedings of the 2017 IEEE Trustcom/BigDataSE/ICESS*. IEEE. 546-553. DOI: 10.1109/Trustcom/BigDataSE/ICESS.2017.283.

[51] Chenli, C., Li, B., Shi, Y., and Jung, T. (2019). Energy-recycling Blockchain with proof-of-deep-learning. In *Proceedings of the 2019*

IEEE International Conference on Blockchain and Cryptocurrency (ICBC). IEEE. 19-23. DOI: 10.1109/BLOC.2019.8751419.

[52] Wasim, M.U., Ibrahim, A.A., Bouvry, P., and Limba, T. (2017). Law as a service (LaaS): Enabling legal protection over a Blockchain network. In *Proceedings of the 2017 14th International Conference on Smart Cities: Improving Quality of Life Using ICT and IoT (HONET-ICT)*. IEEE. 110-114. DOI: 10.1109/HONET.2017.8102214.

[53] Apruzzese, G., Colajanni, M., Ferretti, L., Guido, A., and Marchetti, M. (2018). On the effectiveness of machine and deep learning for cybersecurity. In *Proceedings of the 2018 10th International Conference on Cyber-Conflict (CyCon)*. IEEE. 371-390. DOI: 10.23919/CYCON.2018.8405026.

[54] Wüst, K., and Gervais, A. (2018). Do you need a Blockchain? In *Proceedings of the 2018 Crypto Valley Conference on Blockchain Technology (CVCBT)*. IEEE. 45-54. DOI: 10.1109/CVCBT.2018.00011.

[55] Saritekin, R.A., Karabacak, E., Durgay, Z., and Karaarslan, E. (2018). Blockchain-based secure communication application proposal: Cryptouch. In *Proceedings of the 2018 6th International Symposium on Digital Forensic and Security (ISDFS)*. IEEE. 1-4. DOI: 10.1109/ISDFS.2018.8355380.

[56] Yohan, A., Lo, N.W., and Achawapong, S. (2018). Blockchain-based firmware update framework for internet-of-things environment. In *Proceedings of the International Conference on Information and Knowledge Engineering (IKE)*. WorldCOM. 151-155.

[57] Gupta, Y., Shorey, R., Kulkarni, D., and Tew, J. (2018). The applicability of Blockchain in the Internet of Things. In *Proceedings of the 2018 10th International Conference on Communication Systems and Networks (COMSNETS)*. IEEE. 561-564. DOI: 10.1109/COMSNETS.2018.8328273.

[58] Gu, J., Sun, B., Du, X., Wang, J., Zhuang, Y., and Wang, Z. (2018). Consortium Blockchain-based malware detection in mobile devices. *IEEE Access*, 6, 12118-12128. DOI: 10.1109/ACCESS.2018.2805783.

[59] Chow, S.S., Lai, Z., Liu, C., Lo, E., and Zhao, Y. (2018). Sharding Blockchain. In *Proceedings of the 2018 IEEE International Conference on Internet of Things (iThings) and IEEE Green Computing and Communications (GreenCom) and IEEE Cyber, Physical and Social Computing (CPSCom) and IEEE Smart Data (SmartData)*. IEEE. 1665-1665. DOI: 10.1109/Cybermatics_2018.2018.00277.

[60] Zheng, Z., Xie, S., Dai, H.N., Chen, X., and Wang, H. (2018). Blockchain challenges and opportunities: A survey. *International Journal of Web and Grid Services*, 14(4), 352-375. DOI: 10.1504/IJWGS.2018.095647.

[61] Chawathe, S. (2018). Monitoring Blockchains with self-organizing maps. In *Proceedings of the 2018 17th IEEE International Conference on Trust, Security and Privacy in Computing and Communications/12th IEEE International Conference on Big Data Science and Engineering (TrustCom/BigDataSE).* IEEE. 1870-1875. DOI: 10.1109/TrustCom/BigDataSE.2018.00283.

[62] Tsolakis, A.C., Moschos, I., Votis, K., Ioannidis, D., Dimitrios, T., Pandey, P., ... and García-Castro, R. (2018). A secured and trusted demand response system based on Blockchain technologies. In *Proceedings of the 2018 Innovations in Intelligent Systems and Applications (INISTA).* IEEE. 1-6. DOI: 10.1109/INISTA.2018.8466303.

[63] Signorini, M., Pontecorvi, M., Kanoun, W., and Di Pietro, R. (2018). Bad: Blockchain anomaly detection. *IEEE Access*, 8, 173481-173490. DOI: 10.1109/ACCESS.2020.3025622.

[64] Meng, W., Tischhauser, E.W., Wang, Q., Wang, Y., and Han, J. (2018). When intrusion detection meets Blockchain technology: A review. *IEEE Access*, 6, 10179-10188. DOI: 10.1109/ACCESS.2018.2799854.

[65] Kuo, T.T., and Ohno-Machado, L. (2018). Modelchain: Decentralized privacy-preserving healthcare predictive modeling framework on private Blockchain networks. *arXiv preprint*. arXiv: 1802.01746.

[66] Androulaki, E., Cocco, S., and Ferris, C. (2018). Private and confidential transactions with hyperledger fabric. *IBM Developer Works.*

[67] Singla, K., Bose, J., and Katariya, S. (2018). Machine learning for secure device personalization using Blockchain. In *Proceedings of the 2018 International Conference on Advances in Computing, Communications and Informatics (ICACCI).* IEEE. 67-73. DOI: 10.1109/ICACCI.2018.8554476.

[68] Wang, T. (2018). A unified analytical framework for trustable machine learning and automation running with Blockchain. In *Proceedings of the 2018 IEEE International Conference on Big Data (Big Data).* IEEE. 4974-4983. DOI: 10.1109/BigData.2018.8622262.

[69] Ferrer, E.C., Rudovic, O., Hardjono, T., and Pentland, A. (2018). Robochain: A secure data-sharing framework for human-robot interaction. *arXiv preprint*. arXiv: 1802.04480.

[70] Weng, J., Weng, J., Zhang, J., Li, M., Zhang, Y., and Luo, W. (2019). Deep chain: Auditable and privacy-preserving deep learning with Blockchain-based incentive. *IEEE Transactions on Dependable and Secure Computing*, 18(5), 2438-2455. DOI: 10.1109/TDSC.2019.29 52332.

[71] Chen, R.Y. (2018). A traceability chain algorithm for artificial neural networks using T–S fuzzy cognitive maps in Blockchain. *Future Generation Computer Systems*, 80, 198-210. DOI: 10.1016/j.future.2017. 09.077.

[72] Mylrea, M. (2018). AI-enabled Blockchain smart contracts: Cyber resilient energy infrastructure and IoT. In *Proceedings of the AAAI Spring Symposium Series.*

[73] Regnath, E., and Steinhorst, S. (2018). SmaCoNat: Smart contracts in natural language. In *Proceedings of the 2018 Forum on Specification and Design Languages (FDL).* IEEE. 5-16. DOI: 10.1109/FDL.2018.8524068.

[74] Khraisat, A., Gondal, I., Vamplew, P., and Kamruzzaman, J. (2019). Survey of intrusion detection systems: Techniques, datasets and challenges. *Cybersecurity*, 2(1), 1-22. DOI: 10.1186/s42400-019-0038-7.

[75] Torres, J.M., Comesaña, C.I., and Garcia-Nieto, P.J. (2019). Machine learning techniques applied to cybersecurity. *International Journal of Machine Learning and Cybernetics*, 10(10), 2823-2836. DOI: 2823–2836 (2019). https://doi.org/10.1007/s13042-018-00906-1.

[76] Ismail, L., and Materwala, H. (2019). A review of Blockchain architecture and consensus protocols: Use cases, challenges, and solutions. *Symmetry*, 11(10), 1198. DOI: 10.3390/sym11101198.

[77] Uriarte, R.B., Tiezzi, F., and De Nicola, R. (2019). Dynamic SLAs for clouds, In *Proceedings of the European Conference on Service-Oriented and Cloud Computing*. Eds Aiello, M., Johnsen, E., Dustdar, S., Georgievski, I., Springer, Cham. 34–49. DOI: 10.1007/978-3-319-44482-6_3.

[78] Pillai, A., Sindhu, M., and Lakshmy, K.V. (2019). Securing firmware in Internet of Things using Blockchain. In *Proceedings of the 2019 5th International Conference on Advanced Computing and Communication Systems (ICACCS).* IEEE. 329-334. DOI: 10.1109/ICACCS.2019.8728389.

[79] Alotaibi, B. (2019). Utilizing Blockchain to overcome cybersecurity concerns in the internet of things: A review. *IEEE Sensors Journal*, 19(23), 10953-10971. DOI: 10.1109/JSEN.2019.2935035.

[80] Saad, M., Spaulding, J., Njilla, L., Kamhoua, C., Shetty, S., Nyang, D., and Mohaisen, A. (2019). Exploring the attack surface of Blockchain: A systematic overview. *arXiv preprint*. arXiv: 1904.03487.

[81] Kumari, A., Tanwar, S., Tyagi, S., Kumar, N., Obaidat, M.S., and Rodrigues, J.J. (2019). Fog computing for smart grid systems in the 5G environment: Challenges and solutions. *IEEE Wireless Communications*, 26(3), 47-53. DOI: 10.1109/MWC.2019.1800356.

[82] Doku, R., and Rawat, D. (2019). Pledge: A private ledger-based decentralized data-sharing framework. In *Proceedings of the Spring Simulation Conference (SpringSim)*. IEEE. 1-11. DOI: 10.23919/SpringSim.2019.8732913.

[83] Shen, M., Tang, X., Zhu, L., Du, X., and Guizani, M. (2019). Privacy-preserving support vector machine training over Blockchain-based encrypted IoT data in smart cities. *IEEE Internet of Things Journal*, 6(5), 7702-7712. DOI: 10.1109/JIOT.2019.2901840.

[84] Bore, N.K., Raman, R.K., Markus, I.M., Remy, S.L., Bent, O., Hind, M., ... and Weldemariam, K. (2019). Promoting distributed trust in machine learning and computational simulation. In *Proceedings of the 2019 IEEE International Conference on Blockchain and Cryptocurrency (ICBC)*. IEEE. 311-319. DOI: 10.1109/BLOC.2019.8751423.

[85] Lyu, L., Yu, J., Nandakumar, K., Li, Y., Ma, X., and Jin, J. (2019). Towards fair and decentralized privacy-preserving deep learning with Blockchain. *arXiv preprint.* arXiv: 1906.01167, 1-13.

[86] Wang, J., Wang, L., Yeh, W.C., and Wang, J. (2019). Design and analysis of an effective securing consensus scheme for decentralized Blockchain system. In *Proceedings of the International Conference on Blockchain and Trustworthy Systems*. Eds Zheng, Z., Dai, H.N., Tang, M., Chen, X., Springer, Singapore. 212-225. DOI: 10.1007/978-981-15-2777-7_18.

[87] Qin, P., Guo, J., Shen, B., and Hu, Q. (2019). Towards self-automatable and unambiguous smart contracts: Machine natural language. In *Proceedings of the International Conference on e-Business Engineering*. Eds Zheng, Z., Dai, H.N., Tang, M., Chen, X., Springer, Cham. 479-491. DOI: 10.1007/978-981-15-2777-7_18.

[88] Tara, A., Ivkushkin, K., Butean, A., and Turesson, H. (2019). The evolution of Blockchain virtual machine architecture towards an enterprise usage perspective. In *Proceedings of the Computer Science On-line Conference*. Eds Silhavy, R., Springer, Cham. 370-379. DOI: 10.1007/978-3-030-19807-7_36.

[89] Alpaydin, E. (2020). *Introduction to Machine Learning, fourth edition (Adaptive Computation and Machine Learning series) (fourth edition).* The MIT Press. DOI: 10.1017/S0269888906220745.

[90] Salloum, S.A., Alshurideh, M., Elnagar, A., and Shaalan, K. (2020). Machine learning and deep learning techniques for cybersecurity: A review. In *Proceedings of the Joint European-US Workshop on Applications of Invariance in Computer Vision.* Eds Hassanien, A.E., Azar, A., Gaber, T., Oliva, D., Tolba, F., Springer, Cham. 50-57. DOI: 10.1007/978-3-030-44289-7_5.

[91] Sarker, I.H., Kayes, A.S.M., Badsha, S., Alqahtani, H., Watters, P., and Ng, A. (2020). Cybersecurity data science: An overview from machine learning perspective. *Journal of Big Data*, 7(1), 1-29. DOI: 10.1186/s40537-020-00318-5.

[92] Shaukat, K., Luo, S., Varadharajan, V., Hameed, I.A., Chen, S., Liu, D., and Li, J. (2020). Performance comparison and current challenges of using machine learning techniques in cybersecurity. *Energies*, 13(10), 2509. DOI: 10.3390/en13102509.

[93] Taylor, P.J., Dargahi, T., Dehghantanha, A., Parizi, R.M., and Choo, K.K.R. (2020). A systematic literature review of Blockchain cybersecurity. *Digital Communications and Networks*, 6(2), 147-156. DOI: 10.1016/j.dcan.2019.01.005.

[94] Alharbi, T. (2020). Deployment of Blockchain technology in software-defined networks: A survey. *IEEE Access*, 8, 9146-9156. DOI: 10.1109/ACCESS.2020.2964751.

[95] Kim, T.H., and Reeves, D. (2020). A survey of domain name system vulnerabilities and attacks. *Journal of Surveillance, Security and Safety*, 1(1), 34-60. DOI: 10.20517/jsss.2020.14.

[96] Zheng, Z., Xie, S., Dai, H.N., Chen, W., Chen, X., Weng, J., and Imran, M. (2020). An overview on smart contracts: Challenges, advances and platforms. *Future Generation Computer Systems*, 105, 475-491. DOI: 10.1016/j.future.2019.12.019.

[97] Myadam, N.G., and Patnam, B. (2020). Design and implementation of key exchange mechanisms for software artifacts using Ocean protocol. Masters' thesis, Department of Computer Science, Faculty of Computing, Blekinge Institute of Technology.

[98] Kaur, J., and KR, R.K. (2021). The Recent Trends in Cybersecurity: A Review. *Journal of King Saud University-Computer and Information Sciences.* DOI: 10.1016/j.jksuci.2021.01.018.

[99] Novet, J. (2021). Microsoft's big email hack: What happened, who did it, and why it matters. CNBC. Available at: https://www.cnbc.com/2021/03/09/microsoft-exchange-hack-explained.html [accessed July 23, 2021].

[100] Kilincer, I.F., Ertam, F., and Sengur, A. (2021). Machine learning methods for cybersecurity intrusion detection: Datasets and comparative study. *Computer Networks*, 188, 107840. DOI: 10.1016/j.comnet.2021.107840.

[101] Williams, G. (2021). Bitcoin: A brief price history of the first cryptocurrency. Investing News Network. Available at: https://investingnews.com/daily/tech-investing/Blockchain-investing/bitcoin-price-history/#:%7E:text=Bitcoin%20price%20history%3A%20A%20response,US%24250%20in%20April%202013 [accessed July 30, 2021].

[102] Wani, S., Imthiyas, M., Almohamedh, H., Alhamed, K.M., Almotairi, S., and Gulzar, Y. (2021). Distributed denial of service (DDoS) mitigation using blockchain—A comprehensive insight. *Symmetry*, 13(2), 227. DOI: 10.3390/sym13020227.

[103] Khan, A.A., Khan, M.M., Khan, K.M., Arshad, J., and Ahmad, F. (2021). A Blockchain-based decentralized machine learning framework for collaborative intrusion detection within UAVs. *Computer Networks*, 196, 108217. DOI: 10.1016/j.comnet.2021.108217.

[104] Vainshtein, Y., and Gudes, E. (2021). Use of Blockchain for ensuring data integrity in cloud databases. In *Proceedings of the International Symposium on Cyber Security Cryptography and Machine Learning*. Springer, Cham. 325-335. DOI: 10.1007/978-3-030-78086-9_25.

[105] Kochkarov, A.A., Osipovich, S.D., and Kochkarov, R.A. (2020). Analysis of DDoS attacks on Bitcoin cryptocurrency payment system. *Revista ESPACIOS*, 41(03).

[106] Abu-Nimeh, S., Nappa, D., Wang, X., and Nair, S. (2007). A comparison of machine learning techniques for phishing detection. In *Proceedings of the Anti-Phishing Working Groups 2nd Annual eCrime Researchers' Summit.* ACM. 2, 60-69. DOI: 10.1145/1299015.1299021.

[107] Chen, J., Duan, K., Zhang, R., Zeng, L., and Wang, W. (2018). An AI-based super nodes selection algorithm in Blockchain networks. *arXiv preprint*. arXiv: 1808.00216.

[108] Luong, N.C., Xiong, Z., Wang, P., and Niyato, D. (2018). Optimal auction for edge computing resource management in mobile Blockchain networks: A deep learning approach. In *Proceedings of the 2018 IEEE*

International Conference on Communications (ICC). IEEE. 1-6. DOI: 10.1109/ICC.2018.8422743.

[109] McMillan, R. (2014). The inside story of Mt. Gox, Bitcoin's $460 million disaster. Wired. March, 3. Available at: https://www.wired.com/2014/03/bitcoin-exchange/ [accessed June 29, 2021].

[110] Demertzis, K., Iliadis, L., Tziritas, N., and Kikiras, P. (2020). Anomaly detection via Blockchained deep learning smart contracts in industry 4.0. *Neural Computing and Applications*, 32(23), 17361-17378. DOI: https://doi.org/10.1007/s00521-020-05189-8.

[111] Bravo-Marquez, F., Reeves, S., and Ugarte, M. (2019). Proof-of-learning: A Blockchain consensus mechanism based on machine learning competitions. In *Proceedings of the 2019 IEEE International Conference on Decentralized Applications and Infrastructures (DAPPCON)*. IEEE. 119-124. DOI: 10.1109/DAPPCON.2019.00023.

[112] Wiki, E. (2018). On sharding Blockchains. Available at: https://github.com/ethereum/wiki/wiki/Sharding-FAQs [accessed July 12, 2021].

8

Cyberthreat Real-time Detection Based on an Intelligent Hybrid Network Intrusion Detection System

Said Ouiazzane, Malika Addou, and Fatimazahra Barramou

ASYR Team, Laboratory of Systems Engineering (LaGeS), Hassania School of Public Works (EHTP), Casablanca, Morocco
E-mail: sa.ouiazzane@gmail.com

Abstract

Cybercrime continues to threaten the security of web-connected information systems. Moreover, information is becoming more and more valuable and is the heritage of every size organization operating in all areas of expertise. Therefore, implementing an intrusion detection system is now mandatory to monitor network events and anticipate security incidents that would be costly for the affected organizations. In this work, we have proposed a hybrid, intelligent, cooperative, and distributed network intrusion detection system (NIDS) capable of detecting any type of intrusion that may target modern computer networks. Our system comprises two parts: the first part provides anomaly detection using artificial intelligence and the second part detects known cyber-attacks. Anomaly detection was performed by modeling the network baseline during the normal operation of the network using the decision tree (DT) algorithm applied to the CICIDS2017 dataset. The used dataset underwent pre-processing actions, including eliminating infinite, missing, and duplicate values and reducing dimensionality to retain only the most relevant attributes.

Consequently, the DT recognized normal network traffic with up to 99.9% of accuracy and a very low false alarm rate. In addition, Suricata has been used to detect known cyber-attacks, thanks to its multi-threaded functionality.

The proposed hybrid system recognized deviations from the baseline and performed well in detecting known attacks with a higher accuracy and a low false alarm rate.

Keywords: Intrusion detection, cyberthreat, hybrid NIDS, machine learning.

8.1 Introduction

In the digital transformation era that the world is experiencing, technology is now accessible without precedent. Computer networks are the backbone of the Internet and their use has become essential in everyone's life. Each individual has at least one smartphone to connect to the internet. In addition, the advent of Internet-of-Things (IoT) devices has promoted the expansion and development of computer networks. Along with this evolution and development, the situation has become uncontrollable toward cybercrime and the attack surface is expanding. As a result, hackers are becoming increasingly curious and are constantly exploiting the vulnerabilities of connected devices while undertaking illicit activities in the networks. The information passing through the networks is sensitive and can be infiltrated and captured by cybercriminals. This can range from a network of coordinated botnets to a simple polymorphic worm capable of circumventing existing security policies [1].

Thus, network intrusion detection systems (NIDS) are now an indispensable security brick to monitor the behavior of the networks where they are deployed [2].

Furthermore, traditional NIDS systems are challenging to deploy in modern network environments due to diverse communication architectures, manufacturer policies, technologies, standards, and application-specific services [3]. In particular, zero-day attacks can change their behavior at any time and many times before the patch is developed to bypass implemented NIDS easily. As a result, traditional signature detection-based NIDS (SNIDS) are no longer effective in detecting zero-day cyber-attacks and require regular updates of their signature bases to detect variants of all known intrusions, which is practically impossible!

To detect zero-day intrusions and more advanced and sophisticated cyber-attacks, anomaly detection NIDS (ADNIDS) is now the trend and is gaining more and more attention from cybersecurity researchers [4]. This class of NIDS allows the detection of irregularities in network behavior by recognizing what is normal. However, several challenges are faced by ADNIDS, notably the high number of false alarms they generate in terms of false positives and false negatives. In addition, the lack of reliable datasets capable of representing benign network traffic makes the task even more complex and hinders the development of this type of NIDS. Knowing that testing NIDS on datasets that only include known cyber-attack patterns is no longer sufficient. Furthermore, it is important to test NIDS on datasets that approximate unknown and zero-day attacks. However, generating traffic that can represent unknown and zero-day attacks remains an obstacle for the scientific community to evaluate the effectiveness of NIDS.

Researchers are constantly proposing intrusion detection mechanisms. However, most proposed approaches are based on theoretical architectures that are difficult to implement and based on obsolete test datasets. In addition, the proposed models are more complex and do not exploit the power of open-source detection systems such as Suricata and Snort. Furthermore, detecting intrusions in modern information systems is no longer an easy task and requires more advanced analysis tools. For this purpose, we considered using the multi-agent paradigm to minimize the system's complexity while providing an optimal detection performance.

Moreover, modern intrusion detection systems must recognize all types of malicious attacks with minimal false positive and false negative rates. For this purpose and to build an effective intrusion detection system, conventional NIDS must be integrated with anomaly detection-based NIDS to benefit from the advantages of each category of NIDS. The present work aims at taking advantage of the two categories of NIDS: SNIDS and ADNIDS. Indeed, in this chapter, we will present a hybrid NIDS model that combines both the use of an ADNIDS for anomaly detection and a SNIDS to identify known cyber-attacks based on their signatures. Our model is based on the multi-agent paradigm where autonomous, intelligent, and reactive components called agents communicate and collaborate to detect intrusions in networks.

Moreover, Suricata's open source system is a SNIDS brick in our model, thanks to its multi-threaded functionality unlike single-threaded SNIDS. Furthermore, the ADNIDS brick of our model will be based on modeling the network baseline by modeling the benign traffic of the recent CICIDS2017 dataset. The CICIDS2017 dataset has been optimized by cleaning it and reducing its dimensionality to be consumed by the decision tree algorithm afterward. The decision tree algorithm was chosen as the most suitable technique to model the normal behavior of the network, and this was after conducting a comparative study on machine learning and deep learning algorithms that can model the network baseline in one of our previous works [5].

8.2 Related Works

Intrusion detection systems are the future of cybersecurity to protect computerized infrastructures against cybercrime. Thus, researchers in the scientific community are constantly proposing new mechanisms and approaches to develop this field of expertise.

Several research works have opted for a literature review on current intrusion detection systems. In particular, the author of [6] conducted a state-of-the-art study on new trends and advances in NIDS based mainly on machine learning and deep learning techniques. In addition, the author discussed some methodologies adopted to evaluate NIDS and ensure the best possible selection of learning datasets. The authors of [7]–[11] have proposed NIDS models based on the combination of the two artificial intelligence techniques, which are deep learning and machine learning. These works adopted machine learning and deep learning algorithms to build NIDS models capable of detecting cyber-attacks that violate security policies. In particular, the following combinations were used: non-symmetric deep auto encoder with random forest, sparse auto encoder with SVM, and self-taught learning model based on sparse auto encoder and SVM.

Some researchers [12]–[16] just relied on machine learning techniques alone in their proposed NIDS models. Thus, some machine learning algorithms have been used, namely: ensemble method with optimization using BAT algorithm, fast learning network and particle swarm algorithm, multi-level model based on K-means clustering and random forest, and ensemble machine learning methods with voting algorithm, among others.

Some research works have proposed approaches based only on deep learning. In [17]–[29], researchers have relied on deep learning algorithms in their approaches for intrusion detection. Thus, the following algorithms have been used: recurrent neural network, deep neural network, sparse auto encoder, and multiple layer perception, among others.

It is true that the scientific community has dealt with and continues to study the problem of intrusion detection in modern computer networks. Nevertheless, the proposed models are more complex in architecture and require long learning times. Moreover, most of the works mentioned above have used older datasets, namely the KDD-CUP'99 and NSL-KDD datasets, to test their NIDS models. However, these datasets are obsolete and no longer represent modern network traffic that has changed dramatically in the communication protocol, volume, speed, and behavior. In addition, in some works, the authors used the UNSW-NB15 dataset that is recent compared to the datasets mentioned above (KDD-CUP and NSL-KDD), but, unfortunately, they did not address the training data balancing issue, which led to biased results regarding the classification of the minority classes, namely U2R and R2L.

However, an effective NIDS must detect any type of cyber-attack that may target IT infrastructures. Indeed, a good NIDS must be autonomous, distributed, and intelligent so that it can detect both known and unknown (zero-day) cyber-attacks. Thus, the NIDS must opt for a simple hybrid architecture that employs anomaly detection and signature detection. Anomaly detection will be based on a baseline representing the normal operation of the network traffic. This baseline must be developed based on data representing modern network traffic to recognize normal and, therefore, detect irregularities in network traffic.

With this in mind, this work aims to propose a hybrid, autonomous, and distributed NIDS that will be based on agent technology, artificial intelligence techniques, and a signature-based NIDS with multi-threaded functionality. Our system will combine two types of NIDS: anomaly detection-based NIDS (ADNIDS) and signature-based NIDS (SNIDS). The ADNIDS will be based on modeling the network baseline during its normal operation, using the up-to-date CICIDS2017 dataset after undergoing pre-processing operations such as dataset cleaning, balancing, and extracting relevant attributes.

8.3 The Proposed Approach

In this section, we will discuss our proposed NIDS model. To do so, a reminder of the architecture of the model proposed in the previous work will be presented before presenting the new, improved proposed model.

8.3.1 Overview of the Overall Architecture of the Previously Proposed System

Figure 8.1 presents the architecture of the NIDS model proposed in one of our previous works [30]. Indeed, the model is based on a hybrid architecture employing both a signature-based NIDS and an anomaly-based NIDS. The first SNIDS brick detects known attacks while relying on a signature database describing available attack patterns. Then, the second ADNIDS brick detects irregularities concerning the network baseline. Thus, the combination of the two bricks ensures the detection of any attack that could target IT infrastructures to violate the implemented security policies.

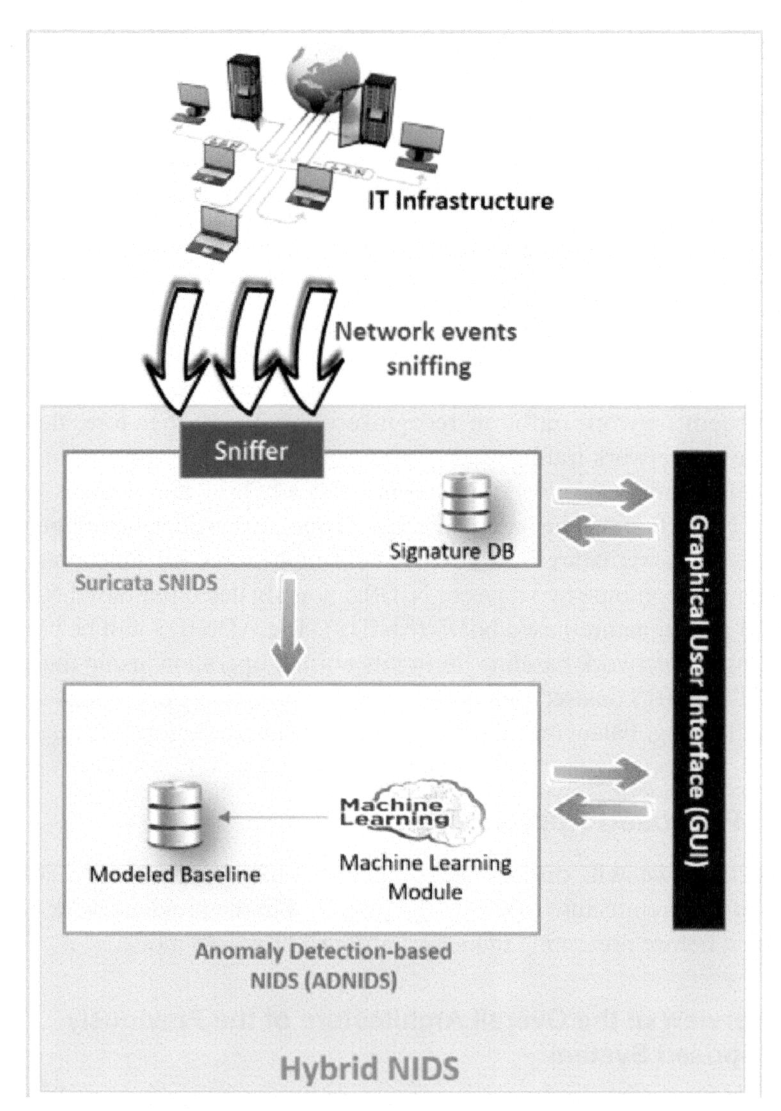

Figure 8.1 Overview of the previous proposed NIDS

The hybrid NIDS model we proposed in [30] is essentially a combination of a SNIDS and an ADNIDS. Suricata ensures the SNIDS function, while the ADNIDS is based on modeling the baseline of the network during its normal functioning. The baseline was built up by modeling the network events present in the CICIDS2017 dataset as normal traffic devoid of intrusions. The idea is to intercept the network packets originating from the IT infrastructure and verify the connection between network event characteristics and the network signature base and baseline.

8.3.2 System Components and Its Operating Principle

Figure 8.2 represents the workflow of the detection mechanism adopted by our previous hybrid NIDS model we proposed in [30]. The hybrid NIDS model relies on combining the SNIDS and ADNIDS bricks to guarantee the detection of any kind of cyberthreats that may threaten the computer data's security principles.

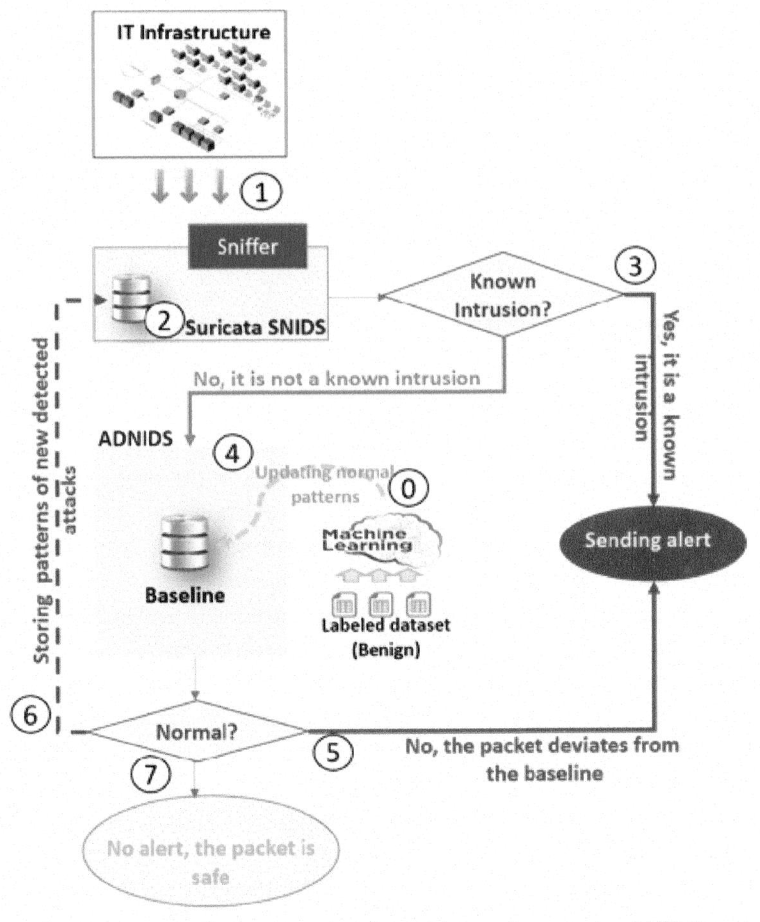

Figure 8.2 The detection workflow of the previous proposed NIDS

Our old NIDS model lists the following components:

- Suricata signature-based NIDS (SNIDS): This component inputs our system and captures network traffic from the IT infrastructure. The SNIDS also detects known attacks based on a signature database that includes the patterns associated with known intrusions.
- Anomaly detection-based NIDS (ADNIDS): This brick detects irregularities concerning the network baseline. ADNIDS is based on a network baseline modeled using machine learning techniques applied on the optimized CICIDS2017 dataset.

For the operating principle of our system, the intrusion detection mechanism is based on the detection of known attacks in a first step before checking the presence of anomalies concerning the network baseline. The SNIDS captures network packets and checks if traces of known attacks are present or not. If it is a known attack, the SNIDS notifies the administrator and if the packet does not correspond to a known attack, the SNIDS forwards the packet in question to the ADNIDS. Thus, ADNIDS checks the packet for irregularities against the network baseline. If the packet is irregular, ADNIDS notifies the administrator and considers it highly likely to be an unknown zero-day attack.

8.3.3 Limitations and Points of Improvement of the Old NIDS Model

The system proposed in our previous work has some limitations and shortcomings that we intend to address through the model that we propose in this work. Indeed, the intrusion detection mechanism in the model, which is the subject of this reminder, relies first on detecting known attacks and then on verifying the deviation from the network baseline. This approach does not consider that normal traffic is much larger than suspect traffic. Thus, the old model processes all incoming network traffic to check for known attack traces and then reprocesses traffic without available attack traces. As a result, both processing phases are done frequently as normal traffic accounts for the largest share while consuming much more resources and causing very long turnaround times. To avoid wasting resources in terms of CPU and RAM throughout the two processing phases, we thought of advancing the brick, ensuring the detection of anomalies compared to the one dedicated to detecting known attacks. In addition, some improvements have been made to the old model to make it more effective and efficient against all types of network attacks.

8.3.4 The Proposed Model Architecture

Figure 8.3 presents the new hybrid NIDS model that we propose after making some improvements and changes to our previous model. Indeed, from a global architecture point of view, the model proposed in this work is based on the combination of two NIDS bricks that are:

- Anomaly detection-based NIDS - ADNIDS: This module is the input of our system and its role is to detect irregularities concerning the network baseline. This ADNIDS is based on a set of intelligent agents that collaborate to carry out anomaly detection missions;
- Signature-based NIDS - SNIDS: This module detects known attacks whose signatures are already recognized by security experts.

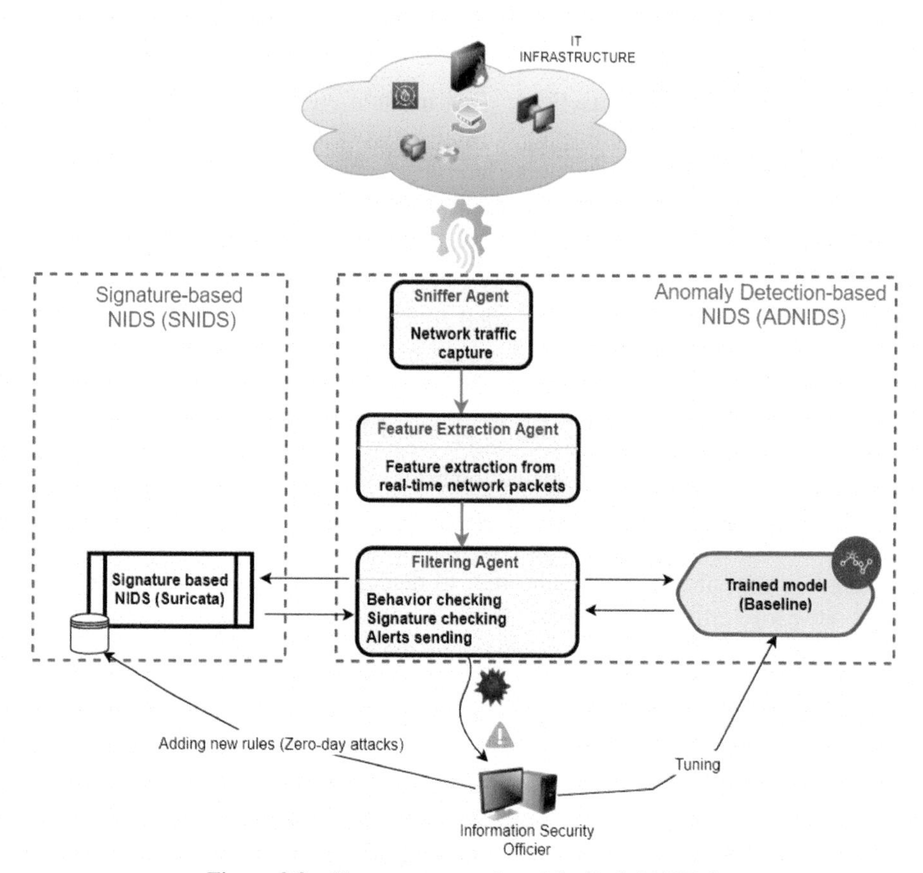

Figure 8.3 The new proposed model of hybrid NIDS

8.3.5 Components of the Proposed New Model

The proposed model is based on the following components:

- ADNIDS: This module is the input component of the entire system and has the following subcomponents:
    - **Sniffer agent (SA):** This agent captures network traffic from the IT infrastructure that requires analysis against network intrusions.
    - **Feature extraction agent (FEA):** This agent receives the network packets captured by the SA and extracts features describing the network traffic behavior.
    - **Filtering agent (FA):** This component receives the network packets with their characteristics extracted by the FEA. Thus, the FA first checks the concordance with the network baseline and then verifies if the packet corresponds to a known intrusion recognized by the SNIDS.
    - **Trained model (baseline):** This module encompasses the set of patterns corresponding to normal network traffic. These patterns are generated using machine learning techniques while training the system on benign traffic from the CICIDS2017 dataset.
- **SNIDS:** This module relies on a signature database to decide whether the captured packet corresponds to a known attack or not. The SNIDS used in this work has multi-threaded functionality, unlike the classical SNIDS based on single-threaded processors.

8.3.6 Operating Principle of the Proposed New Model

Our hybrid model is based on two detection bricks that work together to detect any type of cyber-attacks that could jeopardize current security policies. Indeed, the first brick is based on an ADNIDS comprising a set of intelligent agents that communicate with each other to detect irregularities concerning normal network traffic. The second brick is a SNIDS that detects known intrusions whose signatures are recognized by the classic signature databases. This SNIDS is characterized by its multi-threaded functionality and can perform multiple detection tasks without incurring latency with respect to large network traffic.

Figure 8.4 shows the operating principle of the new proposed hybrid NIDS model while explaining in detail the logic and mechanism adopted for the detection of known and unknown network intrusions. Thus, the intrusion detection mechanism adopted by our system can be described as follows:

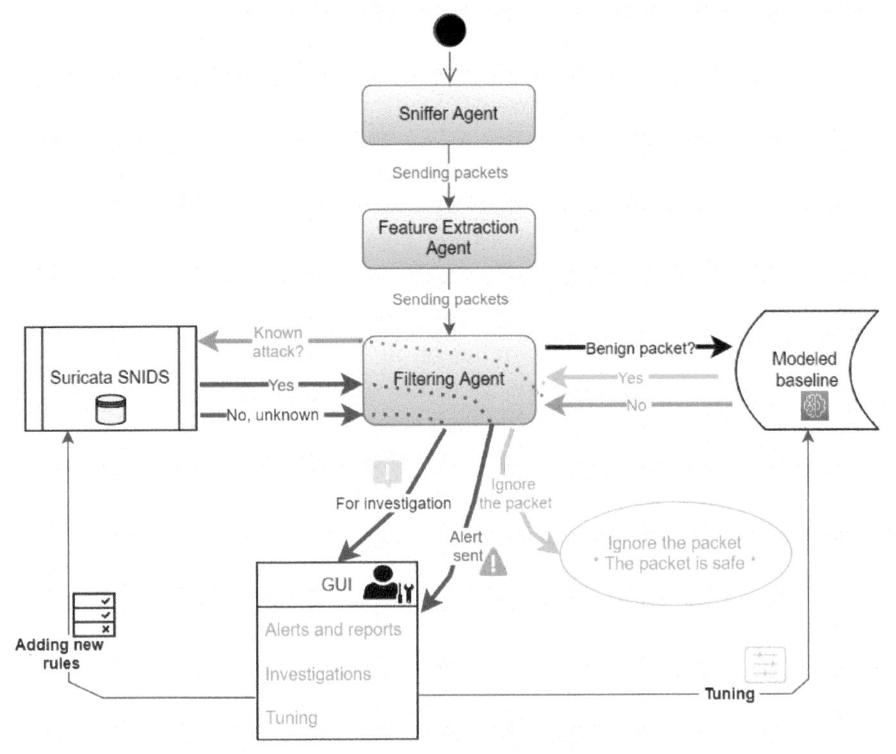

Figure 8.4 The operating principle of the new proposed model of NIDS

- **Step 1:** ADNIDS captures network traffic during this phase to detect whether the network packet is normal or irregular compared to the normal reference profile. To do this, the ADNIDS components perform the following tasks:
  - Real-time capture of network traffic passing through the network via the sniffer agent.
  - Extraction of relevant features and attributes that can describe the behavior of network packets. The feature extraction agent performs the attribute extraction mission.
  - Verification of concordance against the network baseline. This is developed based on the system's training on benign network traffic. The network baseline is built by constructing a model representing the normal operation of the network using a set of machine learning algorithms applied to the CICIDS2017 dataset. Therefore, the filtering agent checks if the network packet characteristics fit into the

normal behavior or present irregularities compared to the normal network operation. At the end of this step, it is considered normal if the packet does not deviate from the baseline. If the packet shows any irregularity concerning the baseline profile, it is suspicious and could be a known attack or an unknown attack. Thus, the filtering agent communicates with the SNIDS to check whether the behavior of the suspicious packet corresponds to a known attack or not.

- **Step 2:** Throughout this phase, the SNIDS decides whether the packet matches a known attack or an unknown behavior that could be a zero-day attack. To do this, the SNIDS relies on its signature database to verify the matching of the characteristics of the suspicious packet against the signature database encompassing the known attack patterns.
- **Step 3:** Administrator notification and baseline tuning.
    - If the suspicious packet matches a known cyber-attack recognized by the SNIDS, the security administrator will automatically notify.
    - If the suspicious packet does not correspond to a known attack, the system considers it an unknown zero-day attack. Thus, the administrator is notified and it is up to him to investigate to confirm the nature of the intrusion in question. Suppose the detected activity is a false positive. In that case, the administrator tunes the baseline and if it corresponds to suspicious activity, the administrator creates a rule in the SNIDS to recognize this activity during future detections.

8.4 Experimentation and Results

In this section, we will just start the experimentation of the ADNIDS brick while addressing the methodology adopted to model the network's baseline and highlighting the results obtained following the classification via the decision tree. To do this, we will first discuss the CICIDS2017 dataset, preprocessing actions in terms of cleaning and dimensionality reduction. Then, we will present the results obtained from the classification of benign traffic using the decision tree. Finally, we will close with discussing the obtained results concerning the modeling of the network baseline.

8.4.1 Modeling the Network Baseline

The system proposed in this work combines two types of NIDS. The first type is the ADNIDS which allows the detection of irregularities in computer

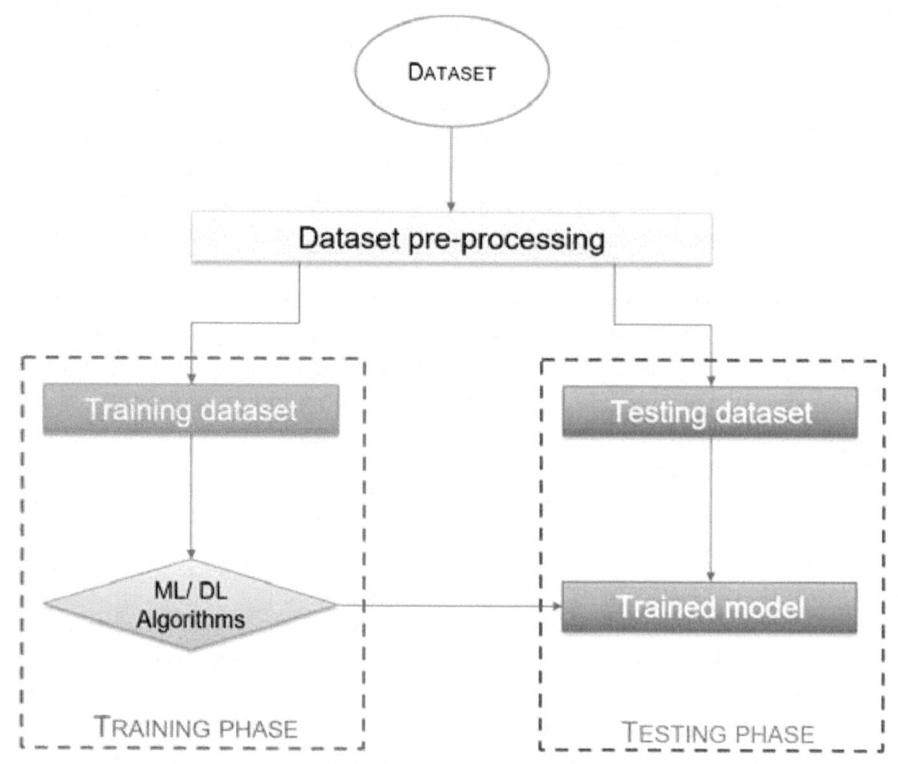

Figure 8.5 The followed approach to model the network baseline

networks while the second type is an SNIDS responsible for recognizing known cyber-attacks.

In this section, we will focus on the ADNIDS brick. Indeed, the ADNIDS brick of our system relies on the network baseline to detect irregularities in the network behaviors. Thus, the modeling of the network baseline relies on the training of the ADNIDS system on benign traffic without any trace of intrusion. Figure 8.5 represents the approach taken to model the network baseline.

To model the network baseline used by ADNIDS, the following steps are taken:

- **Step 1:** Downloading of the dataset, including all benign and malicious network events.
- **Step 2:** Pre-processing the dataset took place to clean it and reduce its dimension in terms of the number of attributes. Thus, the dataset will be

optimized and will help to improve the classification results in terms of performance.

- **Step 3:** Dividing the dataset into two samples: the first for training and the second for testing.
- **Step 4:** Training the ADNIDS on the training dataset based on a set of machine learning and deep learning techniques. At the end of this phase, the classification model is generated and ready to be tested by the test dataset.
- **Step 5:** The testing phase arrives after generating the classification model based on the training dataset. A set of metrics is generated to evaluate the model's performance and see how well this model can recognize benign network traffic. Thus, the test dataset tests the model's effectiveness while giving the model inputs that it has never seen before.
- **Step 6:** After testing the generated models using the different machine learning and deep learning techniques, the best classification algorithm is chosen, based on the metrics of accuracy and false alarm rate. Thus, the best model is the one that has given the best accuracy with a very low false alarm rate and that can recognize the behavior of benign network traffic accurately.

8.4.2 Training Dataset – CICIDS2017

To develop the ADNIDS brick of our hybrid NIDS model, the CICIDS2017 dataset is used to train the ADNIDS brick on benign network traffic devoid of any evidence of intrusions. This dataset contains benign traffic and the most recent cyber-attacks. CICIDS2017 contains updated network packets from modern network traffic and includes data annotated using CICFlowMeter [31]. Thus, we split the CICIDS2017 dataset into two parts: the first part is dedicated to training (80% of the overall dataset) and the second (20%) for testing the modeled baseline pre-processing of the training dataset.

The CICIDS2017 dataset is very recent and truly represents the behavior of modern network traffic. However, some challenges are faced when using this dataset by machine learning and deep learning classification algorithms at training time. Indeed, the CICIDS2017 dataset includes missing and infinite values, which could falsify the results at the time of training. In addition, this dataset contains many attributes that can generate huge learning delays. However, among these attributes, some do not contribute anything to the classification, namely null, correlated, constant, and quasi-constant attributes. Thus, we opted to eliminate the missing and infinite values using the functions of the

Pandas and Numpy libraries. Then, we eliminated constant, quasi-constant, null, and correlated attributes using the Scikit-Learn framework of python.

After pre-processing the CICIDS2017 training dataset, we generated a new optimized dataset with the following characteristics:

- A training dataset without any missing or infinite values.
- A dataset with fewer attributes than the initial one, knowing that the dimensionality of the dataset has been reduced from 79 attributes to 33 relevant attributes. Thus, unnecessary attributes are eliminated and we have kept only the relevant ones that add value at the time of classification.

8.4.3 Classification with the Decision Tree Algorithm

After optimizing the training dataset, it is now time to model the baseline of the network. To do this, the decision tree algorithm is used since this algorithm was chosen as the most suitable in one of our previous works after conducting a comparative study on the use of several algorithms for the classification of benign traffic.

Some metrics are used to evaluate the detection efficiency of the different classes of network traffic, namely:

- *Accuracy*
  Accuracy = (TP + TN)/(TP + FP + FN + TN)
- *Precision*
  Precision = TP/(TP + FP)
- *Recall*
  Recall = TP/(TP + FN)
- *F1-measure*
  F1-Measure = 2 × (Recall × Precision)/(Recall + Precision)

where:

TP → true positives TN → true negatives

FP → false positives FN → false negatives

Table 8.1 highlights the results obtained after applying decision tree as the used learning algorithm. Thus, the evaluation metrics gave very satisfactory results. In particular, the decision tree recognized benign traffic with an accuracy of 99.9% and a very low false alarm rate.

From Table 8.1, which is statistical, the decision tree recognized the benign traffic with an accuracy of 99.91%. This algorithm allowed us to model the network baseline with a low error rate efficiently.

Table 8.1 Obtained statistics after applying decision tree to the optimized CICIDS2017

	Recall	Precision	*F*-Measure	Accuracy
Benign	**0.998**	**0.999**	**0.999**	**99.91 %**
Bot	0.989	0.967	0.978	X
DDoS	1	1	1	X
DoS GoldenEye	0.998	0.992	0.995	X
DoS Hulk	1	1	1	X
DoS SlowHTTPTest	0.994	0.994	0.994	X
DoS Slowloris	0.997	0.997	0.997	X
FTP-Patator	0.999	0.999	0.999	X
Heartbleed	1	0.999	1	X
Infiltration	1	1	1	X
PortScan	1	0.999	0.999	X
SSH-Patator	0.998	0.999	0.999	X
Web Attack Brute Force	0.959	0.98	0.969	X
Web Attack Sql Injection	1	0.999	1	X
Web Attack XSS	0.987	0.961	0.974	X

Table 8.2 Confusion matrix after applying decision tree to the optimized CICIDS2017

	BENIGN	Bot	DDoS	DoS GoldenEye	DoS Hulk	DoS Slowhttptest	DoS slowloris	FTP-Patator	Heartbleed	Infiltration	PortScan	SSH-Patator	WA Brute Force	WA Sql Injection	WA XSS
BENIGN	49927	35	9	10	11	4	1	0	1	0	30	1	5	0	4
Bot	11	1011	0	0	0	0	0	0	0	0	0	0	0	0	0
DDoS	4	0	25685	0	0	0	0	0	0	0	0	0	0	0	0
DoS GoldenEye	0	0	0	2068	5	0	0	0	0	0	0	0	0	0	0
DoS Hulk	8	0	2	6	45963	0	0	0	0	0	4	0	0	0	0
DoS Slowhttptest	3	0	0	0	1	1078	2	0	0	0	0	0	0	0	0
DoS slowloris	0	0	0	1	0	2	1097	0	0	0	0	0	0	0	0
FTP-Patator	1	0	0	0	0	0	0	1603	0	0	0	0	0	0	0
Heartbleed	0	0	0	0	0	0	0	0	1007	0	0	0	0	0	0
Infiltration	0	0	0	0	0	0	0	0	0	989	0	0	0	0	0
PortScan	6	0	0	0	3	0	0	0	0	0	31740	0	1	0	1
SSH-Patator	1	0	0	0	0	0	0	1	0	0	0	1164	0	0	0
WA Brute Force	3	0	0	0	0	0	0	0	0	0	0	0	920	1	35
WA Sql Injection	0	0	0	0	0	0	0	0	0	0	0	0	0	1001	0
WA XSS	0	0	0	0	0	0	0	0	0	0	0	0	13	0	996

Table 8.2 represents the confusion matrix obtained after using the decision tree algorithm for network classification. The results show that the decision tree could recognize the benign traffic more accurately with a very low error rate of false positives and false negatives.

8.4.4 Discussion

Based on the classification results obtained after using the decision tree algorithm to train the ADNIDS brick on benign network traffic, it was obvious that this algorithm can reliably model the network baseline. Thus, the ADNIDS brick of our overall system can now recognize what is normal and, thus, detect irregularities that may be present in the network traffic in real time. In this way, we could model the network baseline that we will implement in the ADNIDS brick of our hybrid NIDS system.

The ADNIDS/SNIDS combination is the next generation of intrusion detection systems to protect modern computer networks against cybercrime. The statistics of the network classification results to model the baseline show that the decision tree is more accurate in recognizing normal network events devoid of any attack traces. Additionally, the use of Suricata as a SNIDS brick has allowed us to take advantage of the strengths of open-source solutions with multi-threaded functionality. The ADNIDS brick efficiently recognized normal network traffic with an accuracy of 99.91%, allowing for more reliable detection of irregularities in deviation from the baseline. Moreover, Suricata is a SNIDS with a huge community that keeps improving its recognition of known cyber-attacks and its detection performance. As such, the effectiveness of the entire NIDS system depends on the effectiveness of both the SNIDS and ADNIDS bricks.

8.5 Conclusion

In this work, we have proposed a hybrid NIDS model that takes advantage of the benefits offered by ADNIDS and SNIDS. Our model combines ADNIDS for irregularity detection and SNIDS for known attack recognition while opting for an agent-based architecture.

The ADNIDS brick is based on modeling the network baseline using the more recent CICIDS2017 dataset. Thus, the training dataset is clean pre-processed to make the classification results as reliable as possible.

The work is not yet complete here and the following tasks remain to be done:

- Implementation and testing of the Suricata SNIDS
- Testing and validation of the entire system in a real computer network

References

[1] Sohi S et al., "RNNIDS: Enhancing network intrusion detection systems through deep learning." *Comput.* Security. 2021.

[2] Ouiazzane S, Addou M and Barramou F, "A Multiagent and Machine Learning Based Denial of Service Intrusion Detection System for Drone Networks." Springer, 2022.

[3] Subbarayalu et al., "Hybrid network intrusion detection system for smart environments based on internet of things." *Comput. J.* 2019.

[4] Ouiazzane S, Barramou F, and Addou M, "Towards a multi-agent based network intrusion detection system for a flee)." http://dx.doi.org/10.14569/IJACSA.2020.0111044

[5] Ouiazzane S, Addou M, and Barramou F, "A multiagent and machine learning based hybrid NIDS for known and unknown cyber-attacks." Int. J. Adv. Comput. Sci. Appl. 2021;12(8). http://dx.doi.org/10.14569/IJACSA.2021.0120843

[6] Zeeshan A et al., "Network intrusion detection system: A systematic study of machine learning and deep learning approaches," 2020.

[7] Shone N et al., "A deep learning approach to network intrusion detection." *IEEE Trans. Emerg. Top Comput.* Intell. 2018;2(1):41-50. https://doi.org/10.1109/TETCI.2017.2772792.

[8] Yan B et al., "Effective feature extraction via stacked sparse autoencoder to improve intrusion detection system." *IEEE Access*. 2018;6:41238-41248. https://doi.org/10.1109/ACCESS.2018.2858277.

[9] Naseer S et al., "Enhanced network anomaly detection based on deep neural networks." *IEEE Access*. 2018;6:48231-48246. https://doi.org/10.1109/ACCESS.2018.2863036.

[10] Al-Qatf M et al., "Deep learning approach combining sparse autoencoder with SVM for network intrusion detection." *IEEE Access*. 2018;6:52843-52856. https://doi.org/10.1109/ACCESS.2018.2869577.

[11] Marir N et al., "Distributed abnormal behavior detection approach based on deep belief network and ensemble SVM using spark." *IEEE Access*. 2018;6:59657-59671. https://doi.org/10.1109/ACCESS.2018.2875045.

[12] Shen et al., "An ensemble method based on selection using bat algorithm for intrusion detection." 2018;61(4):526-538. https://doi.org/10.1093/comjnl/bxx101.

[13] Ali MH et al., "A new intrusion detection system based on fast learning network and particles warm optimization." *IEEE Access*. 2018;6:20255-20261. https://doi.org/10.1109/ACCESS.2018.2820092.

[14] Yao H et al., "MSML: A novel multilevel semi-supervised machine learning framework for intrusion detection system." *IEEE IoT J.* 2018;6(2):1949-1959. https://doi.org/10.1109/JIOT.2018.2873125.

[15] Gao X et al., "An adaptive ensemble machine learning model for intrusion detection." *IEEE Access.* 2019;7:82512-82521. https://doi.org/10.1109/ACCESS.2019.2923640.

[16] Karatas G et al., "Increasing the performance of machine learning-based IDSs on an imbalanced and up-to-date dataset." *IEEE Access.* 2020;8:32150-32162. https://doi.org/10.1109/ACCESS.2020.2973219.

[17] Yin C et al., "A deep learning approach for intrusion detection using recurrent neural networks." *IEEE Access.* 2017;5:21954-21961. https://doi.org/10.1109/ACCESS.2017.2762418.

[18] Jia Y et al., "Network intrusion detection algorithm based on deep neural network." *IET Inf.* Secur. 2018;13(1):48-53. https://doi.org/10.1049/iet-ifs.2018.5258.

[19] Wang Z et al., "Deep learning-based intrusion detection with adversaries." *IEEE Access.* 2018;6:38367-38384. https://doi.org/10.1109/ACCESS.2018.2854599.

[20] Xu C et al., "An intrusion detection system using a deep neural network with gated recurrent units." *IEEE Access.* 2018;6:48697-48707. https://doi.org/10.1109/ACCESS.2018.2867564.

[21] Papamartzivanos et al., "Introducing deep learning self-adaptive misuse network intrusion detection systems." *IEEE Access.* 2019;7:13546-13560. https://doi.org/10.1109/ACCESS.2019.2893871.

[22] Khan FA et al., "A novel two-stage deep learning model for efficient network intrusion detection." *IEEE Access.* 2019;7:30373-30385. https://doi.org/10.1109/ACCESS.2019.2899721.

[23] Xiao Y et al., "An intrusion detection model based on feature reduction and convolutional neural networks." *IEEE Access.* 2019;7:42210-42219. https://doi.org/10.1109/ACCESS.2019.2904620.

[24] Vinayakumar R et al., "Deep learning approach for intelligent intrusiondetection system." *IEEE Access.* 2019;7:41525-41550. https://doi.org/10.1109/ACCESS.2019.2895334.

[25] Wei P et al., "An optimization method for intrusion detection classification model based on deep belief network." *IEEE Access.* 2019;7:87593-87605. https://doi.org/10.1109/ACCESS.2019.2925828.

[26] Malaiya RK et al., "An empirical evaluation of deep learning for network anomaly detection." *IEEE Access.* 2019;7:140806-140817. https://doi.org/10.1109/ACCESS.2019.2943249.

[27] Jiang K et al., "Detection combined hybrid sampling with deep hierarchical network." *IEEE Access*. 2020;8:32464-32476. https://doi.org/10.1109/ACCESS.2020.2973730.

[28] Yang Y et al., "Network intrusion detection based on supervised adversarial variational auto-encoder with regularization." *IEEE Access*. 2020;8:42169-42184. https://doi.org/10.1109/ACCESS.2020.2977007.

[29] Yu Y et al., "An intrusion detection method using few-shot learning." *IEEE Access*. 2020;8:49730-49740. https://doi.org/10.1109/ACCESS.2020.2980136.

[30] Ouiazzane S, Addou M, and Barramou F, (2022), A Suricata and Machine Learning Based Hybrid Network Intrusion Detection System. In: Maleh Y, Alazab M, Gherabi N, Tawalbeh L, Abd El-Latif AA (eds), Advances in Information, Communication and Cybersecurity. ICI2C 2021. Lecture Notes in Networks and Systems, vol. 357. Springer, Cham. https://doi.org/10.1007/978-3-030-91738-8_43

[31] Ouiazzane S et al., "A Multiagent and Machine Learning Based Denial of Service Intrusion Detection System for Drone Networks." Springer, 2022.

9

Intelligent Malware Detection and Classification using Boosted Tree Learning Paradigm

S. Abijah Roseline and S. Geetha

School of Computer Science and Engineering, Vellore Institute of Technology - Chennai Campus, India
E-mail: 1abijahroseline.s2017@vitstudent.ac.in; geetha.s@vit.ac.in

Abstract

Most business processes are now taking place in the cloud, which has never been the case before. Cost management, increased flexibility, improved productivity, and better employee experiences are just a few of the advantages of digital transformation. However, it also introduces new security issues. Cybercriminals continually employ additional features such as code obfuscation to create malware versions and avoid detection by traditional malware detection tools. In this chapter, we present a robust PE-based malware detection and classification system based on boosted tree ensemble learning paradigm and a comparison of the performance of the proposed approach versus other tree models and stacked tree models. Ensemble learning provides a structured approach to combining the predictive capability of various learners. A comprehensive series of tests are conducted to investigate the proposed methodology with two representative malware datasets, such as ClaMP and BIG2015 malware datasets. The effectiveness of using static header-based, content-based, and structure-based features to detect PE malware is explored. The proposed MDS addresses malware detection concerns by increasing accuracy, reducing processing time, removing obfuscation, and detecting new, zero-day malware.

Keywords: Malware detection, cyber security, tree-based methods, ensemble methods, stacked ensembles, machine learning.

9.1 Introduction

Malware is a malicious software created to destroy or exploit devices or computer systems. Malware of various forms, such as viruses, worms, Trojan horses, ransomware, and spyware, continues to be a severe concern for enterprises and governmental organizations. Anti-virus signatures, heuristics, and behavioral patterns in virtual environments are used in traditional malware detection systems, requiring a significant amount of human effort from security professionals and researchers. The signature-based approach involves a collection of manually generated rules which strives to recognize various classes of known malware samples. These rules are standard and fragile, unable to detect new malware even if it leverages the same behavior.

Organizations regularly struggle to keep up with malware threats as new attacks and versions appear. Machine learning (ML) techniques can detect unknown and zero-day malware by automatically recognizing malware patterns based on enormous volumes of existing data. Contrary to heuristic and signature-based techniques, this unique feature has made ML an essential aspect of a contemporary malware detection solution.

Windows executables, object code, and dynamic link libraries (DLLs) all employ the portable executable (PE) file format. Malware attackers make extensive use of this file type. Figure 9.1 depicts the five-year statistics [1] for Windows and Android malware from 2017 to 2021. When compared to Android malware, the Windows platform has the most infections. In 2021, there are 107.5 million samples of Windows malware discovered, according to a recent threat analysis. In 2017, there were 6.2 million Android malware samples, which have declined to 2.96 million in 2021. During the last five years, the number of Windows malware has grown, while Android malware has fallen. This indicates that cybercriminals are more interested in focusing on Windows malware.

After being trained with millions of malicious and clean PE samples, the proposed boosted tree-based ML classifier can identify malware patterns in raw bytes. Tree-based classification models are a form of supervised machine learning method that splits training data into subsets using a set of conditional statements. Each succeeding partition increases the model's complexity, which may be used to build a predictive model. The final model can be represented as a series of logical assertions that define the dataset.

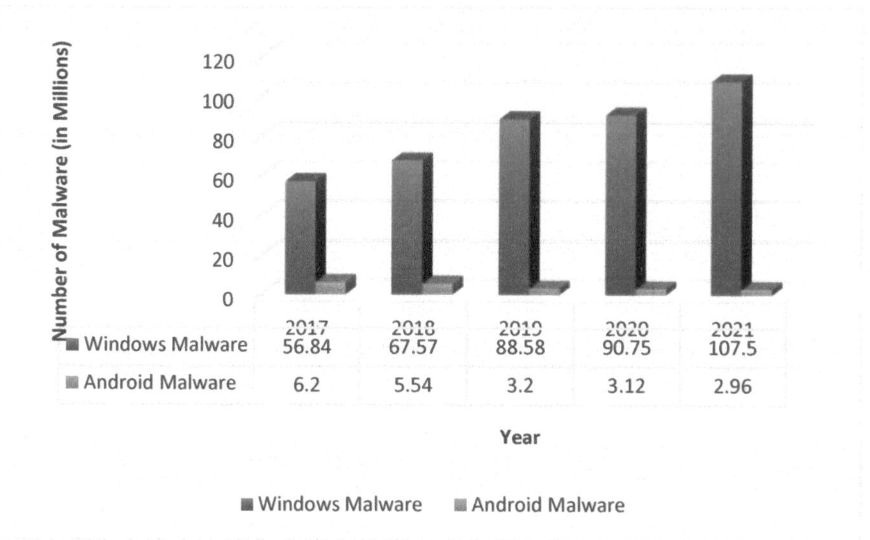

Figure 9.1 Malware count statistics for windows and android platforms from 2017 to 2021

Decision trees are appropriate for modest datasets because they are simple to construct and even simpler to learn.

This chapter proposes an effective, non-stacked, and boosted ensemble malware detection approach to detect and classify malware and its variants. Stacking is a machine learning ensemble approach that trains the MDS by aggregating the predictions from various high-performing tree-based machine learning algorithms. The proposed MDS makes use of the extreme gradient boosting (XGB) model to detect malware, and performance evaluations are carried out by analyzing tree-based ML algorithms such as decision tree (DT), random forest (RF), extra trees (ET), and stacked models (stacked DT, stacked RF, stacked ET, and stacked XGB). The proposed boosted tree-based MDS performance is assessed using two standard datasets, with results demonstrating good accuracy in detecting malware.

The contributions of this chapter include the following:

- Proposed a non-stacked, boosted tree-based methodology to perform binary and multiclass PE malware classification.
- Other tree models and stacked tree models are used for performance assessment on two malware datasets: ClaMP and BIG2015.
- Explored various types of PE file features such as header features, content features, and structural features.

The rest of the chapter is organized as follows. Section 2 provides a survey on the existing malware detection systems. Section 3 explains the proposed malware detection method. Section 4 discusses the experimental results and evaluations for the two PE malware datasets. Section 5 summarizes the chapter.

9.2 Literature Survey

This section covers PE malware detection and classification that uses static analysis, dynamic analysis, vision-based analysis methods, and hybrid analysis methods. Machine learning algorithms have been designed and validated using static or dynamic or vision-based features extracted from program files.

Static analysis methods extract static features such as function length frequency, function call graph, byte sequences, string sequences, opcode sequences, and PE file features. Wadkar *et al.* [2] proposed a malware evolution detection approach using support vector machine (SVM) by extracting static features from PE binaries. Their method required tracking the SVM weight updates over time using statistics. Their technique identified different timepoints when a significant change occurred.

N-grams, opcode sequences, API call sequences, system call sequences, and control flow graphs are features extracted by dynamic analysis methods. Du *et al.* [3] presented a dynamic malware detection system called Magpie, which exploits behavior features like API call parameters and API traces. The API call sequences construct classified behavior graphs to identify malware versions. Their system is not general, and it is incapable of detecting sophisticated malware. It has the same intrinsic disadvantage as all dynamic approaches. Ding *et al.* [4] developed a dynamic taint analysis model to obtain system call parameters using taint tags. Malware behaviors are depicted as dependency graphs based on system calls. They used a weighted graph to capture the common behavior of a malware class, and they devised a behavior matching approach for malware detection and classification. Like other dynamic detection systems, their system has drawbacks, such as tracking only a subset of executable activities and being useless if malware hides its risky activities when executing in a safe setting. Azeez *et al.* [5] presented an ensemble-learning-based method in which the first phase uses neural networks, and 15 machine learning models were used as final-phase classifiers. Their technique is effective against known malware but ineffective against new, zero-day malware. They did not employ large datasets for analysis.

Vision-based techniques have recently been used in the detection of PE malware executables. The binaries are visualized as grayscale images and vision-based features are extracted. The executables are analyzed by visualizing them as images. The executable is then characterized using image-based features. This involves either considering the whole image as a single feature vector or identifying the local and global features of the binary image. Malware classification is carried out by leveraging machine learning [13], [15] or deep learning [14], [16], [17] techniques. Nataraj *et al.* [6] presented a malware detection system that extracted texture features from grayscale images of executables and classified malware using K-nearest neighbors (KNN) and Euclidean distance. Their approach was less computationally intensive than the n-gram method. Fu *et al.* [7] employed a combination of global and local features to perform malware classification. The global features such as texture and color features were obtained from RGB color images of PE binaries. Their model also extracted local features from code and data sections of PE files.

9.3 The Proposed Methodology

The proposed model detects and classifies malware using a non-stacked boosted tree-based detection approach, and its efficiency is evaluated by comparing the proposed MDS to both non-stacked and stacked tree-based models. Non-stacked tree-based models like DT, RF, ET, and XGB, as well as stacked tree models like stacked DT, stacked RF, stacked ET, and stacked XGB, were utilized to compare the performance of the proposed boosting-based MDS. The stacked tree-based model is the meta-learner that takes the training data and runs it through multiple base learners such as DT, RF, ET, and XGB. Then, the predictions from these models are generated.

9.3.1 The Rationale for the Choice of Boosting Classifier

In boosting, trees are constructed sequentially, with each succeeding tree attempting to minimize the errors of the preceding three. Each tree builds on the knowledge of its precedents and updates the residual errors. As a result, the following tree in the sequence will learn from an updated version of the residuals. The base learners in boosting are weak learners with a strong bias and predictive ability that is slightly better than random guessing. Each of these base learners provides crucial information for prediction, allowing the boosting strategy to combine these weak learners to build a strong learner efficiently. The ultimate powerful learner reduces both the bias and the variance. Boosting, instead of bagging approaches such as random forest,

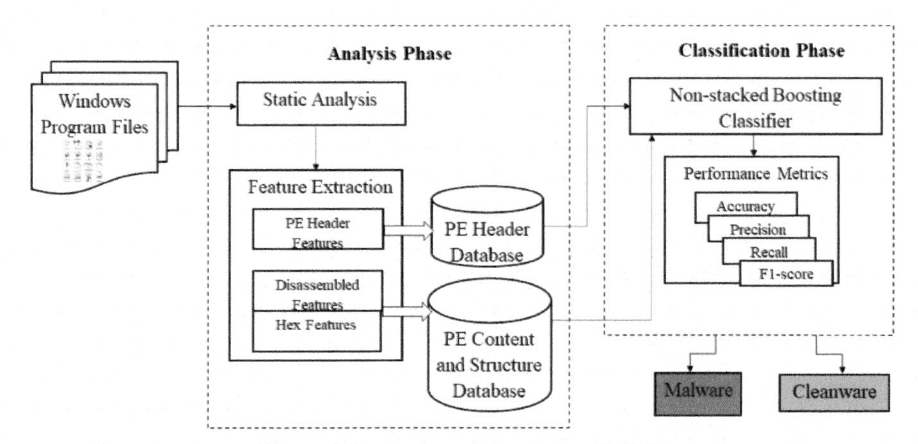

Figure 9.2 Design of the proposed MDS

in which trees are generated to their greatest degree, uses trees with minimal splits. Such little, shallow trees are straightforward to interpret.

9.3.2 Overview

Figure 9.2 shows the overall architecture of the proposed malware detection system, which includes two phases, namely, the analysis phase and the classification phase. The Windows PE files are analyzed to obtain static PE file features in different dimensions. The header features are stored in the PE header database. In contrast, the disassembled file features and hex features are stored in the PE content and structure database to perform efficient malware detection and classification. In the classification phase, the proposed non-stacked boosting-based tree model detects and classifies malware using the different types of extracted PE file features. The performance metrics such as accuracy, precision, recall, and $f1$-score are assessed in discriminating malware from clean ware binaries.

9.3.3 Classifiers used for Evaluation

9.3.3.1 Decision Tree (DT)

A decision tree [8] is a useful approach to represent data because it follows a tree-like arrangement and considers all of the different paths that can lead to the final result. It usually begins with a single node and branches out into different outcomes. Decision-tree classifiers might produce ridiculously complicated trees that do not generalize well to new malware. This is referred to as overfitting. Decision trees can be volatile because minor changes in the data can result in the generation of an entirely new tree. This is known as variance, and it must be reduced using techniques such as bagging and boosting. When some classes are dominant, decision tree classifiers produce unbalanced trees.

9.3.3.2 Random Forest (RF)

The random forest [9] classifier is a bagging approach that uses both bagging and feature randomization to produce an independent forest of decision trees. Feature randomization, also called bagging, creates random feature subsets, resulting in decision trees with weak correlation. Random forests choose only a subset of the potential feature splits, whereas decision trees examine all potential feature splits.

9.3.3.3 Extra Trees Classifier (ET)

Extra trees [10] or extremely randomized trees is similar to random forest in that it generates several trees and divides nodes using random feature subsets. Still, it differs in two major ways: it does not bootstrap data (meaning it samples without replacement) and nodes are divided into randomized splits rather than optimal splits. Randomness in extra trees occurs from the random partitions of all data, rather than from data bootstrapping.

9.3.3.4 XGBoost

XGBoost [11], also known as extreme gradient boosting, is an efficient and scalable algorithm that promotes quick learning via parallel processing computing and effectively uses memory.

9.3.3.5 Stacked Ensembles

The stacked ensemble model uses four base learners: DT, RF, ET, and XGB. The stacked model uses DT as the predictive model and is termed as stacked DT. When RF is used as the predictive model, it is called stacked RF. When ET is used as the predictive model, it is called stacked ET. When XGB is used as the predictive model, it is termed stacked XGB.

9.4 Experimental Results

9.4.1 Datasets

The efficiency of the proposed model was evaluated using two datasets such as ClaMP (classification of malware with PE headers) and Kaggle malware

Table 9.1 Dataset details of BIG2015 malware dataset

S. No	Class	Type	Number of samples
1	Ramnit	Worm	1541
2	Lollipop	Adware	2478
3	Kelihos_ver3	Backdoor	2942
4	Vundo	Trojan	475
5	Simda	Backdoor	42
6	Tracur	TrojanDownloader	751
7	Kelihos_ver1	Backdoor	398
8	Obfuscator.ACY	Any obfuscated malware	1228
9	Gatak	Backdoor	1013

datasets. The first dataset, ClaMP [18], is a binary malware dataset with 5184 samples containing two classes: malware (2683 samples) and cleanware (2501 samples). The dataset includes 55 features extracted from PE binary files. The second dataset, the Kaggle BIG2015 malware [12] dataset, is a multiclass malware dataset that includes 10,868 samples containing nine malware classes with 1804 PE file features. Table 9.1 provides the classes, their type, and the number of samples contained in each class of the BIG2015 malware dataset.

9.4.1.1 Features of ClaMP Malware Dataset

The ClaMP malware dataset includes features extracted from all three main PE headers: IMAGE_DOS_HEADER, FILE_HEADER, and OPTIONAL_HEADER. The IMAGE_DOS_HEADER has 19 features, the FILE_HEADER has 7 features, and OPTIONAL_HEADER has 29 features.

- **IMAGE_DOS_HEADER:**
  - e_magic, e_cblp, e_cp, e_crlc, e_cparhdr, e_minalloc, e_maxalloc, e_ss, e_sp, e_csum, e_ip, e_cs, e_lfarlc, e_ovno, e_res, e_oemid, e_oeminfo, e_res2, and e_lfanew
- **FILE_HEADER:**
  - Machine, NumberOfSections, CreationYear, PointerToSymbol-Table, NumberOfSymbols, SizeOfOptionalHeader, and Characteristics
- **OPTIONAL_HEADER:**
  - Magic, MajorLinkerVersion, MinorLinkerVersion, SizeOfCode, SizeOfInitializedData, SizeOfUninitializedData, AddressOfEntry Point, BaseOfCode, BaseOfData, ImageBase, SectionAlignment,

FileAlignment, MajorOperatingSystemVersion, MinorOperating SystemVersion, MajorImageVersion, MinorImageVersion, Major-SubsystemVersion, MinorSubsystemVersion, SizeOfImage, Size-OfHeaders, CheckSum, Subsystem, DllCharacteristics, SizeOf-StackReserve, SizeOfStackCommit, SizeOfHeapReserve, Size-OfHeapCommit, LoaderFlags, and NumberOfRvaAndSizes

9.4.1.2 Features of BIG2015 Malware Dataset

The BIG2015 malware dataset [12] contains 1805 hybrid malware features (both content-based and structural features) extracted from the hex and assembly views. The Hex-dump-based features include *n*-gram, metadata, entropy, image representation, and string length. The disassembled file features include metadata, symbol, operation code, register, application programming interface (API), section, data define, and frequency of some miscellaneous keywords (MISC). The dataset does not include PE header features.

The category of features based on hex file includes *n*-gram, metadata, entropy, image representation, and string length. To acquire useful information regarding the type of malware, the representation of a malware sample as a series of hex values can be efficiently expressed using *n*-gram analysis. The byte sequence is represented by the *n*-gram features, which are characterized with a vector of 256 dimensions. The metadata (MD1) includes the file size and the address of the initial bytes sequence. The address is a hexadecimal number that is transformed to the decimal equivalent for consistency with the other feature variables. Entropy (ENT) is a metric of disorder in the distribution of bytes in bytecode, with a value ranging from 0 (order) to 8 (randomness) that can be used to determine whether or not obfuscation is present. The features are calculated using the sliding window method and statistics of entropy sequences such as quantiles, percentiles, mean, and variance of the obtained distribution. Furthermore, the entropy of all bytes in malware is estimated. Each malware can be represented as an image and the features extracted include the texture features such as Haralick features and Local Binary Patterns. The use of string features is impractical since a large number of junk values are retrieved with real strings. To minimize noise and prevent overfitting, only histograms of string length distribution (STR) are employed.

The extracted features from disassembled files include metadata, symbol, operation code, register, application programming interface (API), section, data define, and miscellaneous features. The metadata (MD2) list includes

Table 9.2 Some important PE malware features in the BIG2015 malware dataset

Category	Features
Section (SEC)	section_names .bss
	section_names_.data
	section_names _edata
	section_names _.idata
	section_names _rdata
	section_names_.rsrc
	section_names_.text
	section_names_.tls
	section_names_.reloc
	Num_Sections
	Unknown_Sections
	Unknown_Sections_lines
	Known_Sections_por
	Unknown_Sections_por
	Unknown_Sections_lines_por
	.text_por
	.data_por
	.bss_por
	.rdata_por
	.edata_por
	.idata_por
	.rsrc_por
	.tls_por
	.reloc_por
Data Define (DP)	db_por
	dd por
	dw_por
	dc_por
	db0_por
	dbN0_por
	dd_text
	db_text
	dd_rdata
	db3_rdata
	db3_data
	db3_all
	dd4
	dd5
	dd6
	dd4_all
	dd5_all

Table 9.2 Continued.

Category	Features
	dd6_all db3_idata db3_NdNt dd4_NdNt dd5_NdNt

features such as file size and the number of lines in each file examined after disassembly. The dataset considers the frequencies of the symbols (SYM), -, +, *,], [,?, @, as a high frequency of these characters indicates code meant to elude detection. The assembly language instructions are represented using operation codes (OPC). The frequencies of selected operation codes are measured for each sample. The frequency of using registers (REG) and using APIs are included. The section (SEC) features include .text, .data, .bss, .rdata, .edata, .idata, .rsrc, .tls, and .reloc. These section features are critical for efficiently classifying malware. The data-defined features are largely db, dw, and dd instructions, which are used to set the byte, word, and doubleword values, respectively. The frequency of manual keywords (MISC) is extracted from the disassembled file. Some significant features suitable for classification are shown in Table 9.2.

9.5 Results and Discussion

The experiments were performed on Windows 64-bit operating system, x64-based Intel(R) Core(TM) i5-1135G7 @ 2.40-GHz processor. The proposed malware detection system was implemented using the Python programming language, which included all essential packages. The proposed boosting-based MDS and various single and stacked tree-based MDS performance is compared for two PE malware datasets. The results are shown in Table 9.3. The proposed MDS performed better on the ClaMP malware dataset, with an accuracy of 98.65%, precision of 0.9865, recall of 0.9865, and $F1$-score of 0.9865. The proposed model performed well on the BIG2015 malware dataset, with 99.86% accuracy, 0.9986 precision, 0.9986 recall, and 0.9986 $F1$-score. The stacked models did not outperform the non-stacked XGB model in detecting malware.

The confusion matrices for the proposed MDS with ClaMP and BIG2015 malware datasets are shown in Figures 9.3 and 9.4. Since the ClaMP dataset is a binary dataset, the confusion matrix is a 2×2 matrix, whereas the

Table 9.3 Comparison of the proposed MDS with other tree-based MDS for the two PE malware datasets

Models	Clamp dataset				BIG2015 malware dataset			
	Accuracy (%)	Precision	Recall	*F*1-score	Accuracy (%)	Precision	Recall	*F*1-score
DT	96.05	0.9605	0.9605	0.9605	99.13	0.9912	0.9913	0.9912
RF	97.78	0.9778	0.9778	0.9778	99.68	0.9968	0.9968	0.9968
ET	98.17	0.9817	0.9817	0.9817	99.40	0.9941	0.9940	0.9940
Stacked DT	96.05	0.9605	0.9605	0.9605	99.08	0.9909	0.9908	0.9908
Stacked RF	98.46	0.9846	0.9846	0.9846	99.68	0.9968	0.9968	0.9968
Stacked ET	98.46	0.9846	0.9846	0.9846	99.68	0.9968	0.9968	0.9968
Stacked XGB	96.05	0.9605	0.9605	0.9605	99.13	0.9912	0.9913	0.9912
XGB	**98.65**	**0.9865**	**0.9865**	**0.9865**	**99.86**	**0.9986**	**0.9986**	**0.9986**

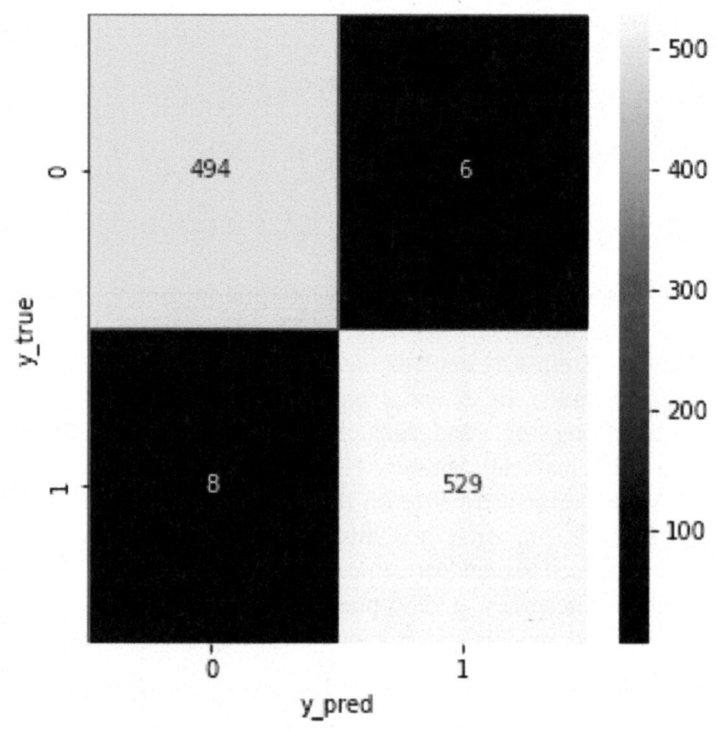

Figure 9.3 Confusion matrix of the proposed MDS for ClaMP malware dataset

multiclass BIG2015 dataset is a 9×9 matrix. The diagonal elements of both confusion matrices indicate that the proposed non-stacked boosting model made the correct predictions. For the ClaMP malware dataset, the number of cleanware samples correctly classified as cleanware by the proposed MDS is 494. The number of malware samples correctly classified as malware is 529. The number of cleanware samples incorrectly classified as malware is 8. The number of malware samples incorrectly classified as cleanware is 6. For the BIG2015 malware dataset, the predictions of various classes by the proposed MDS are represented in Figure 9.4. The non-diagonal elements show the incorrect classifications by the proposed model, which is very few. This implies that the proposed MDS is significantly more effective in distinguishing between malware and cleanware and malware variants.

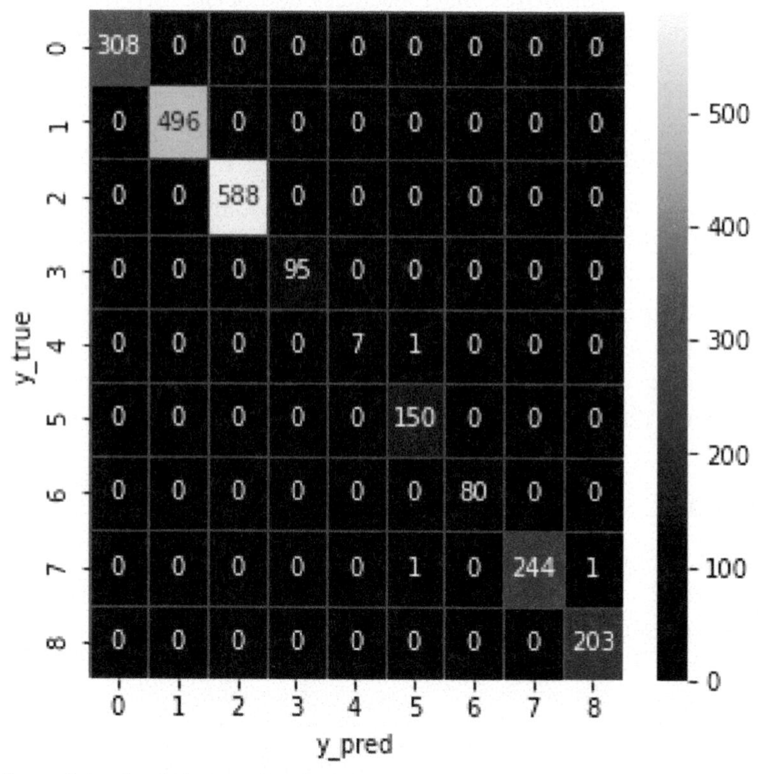

Figure 9.4 Confusion matrix of the proposed MDS for BIG2015 malware dataset

9.6 Conclusion

The proposed malware detection system uses static PE features such as header features and content-based and structural features to discriminate malware from cleanware files. The boosted classifier approach is used for malware detection and classification. A complete series of tests are carried out using stacked and non-stacked tree-based ML models to assess the proposed MDS, and the results are compared. The proposed tree-based MDS outperformed the other tree-based methods with classification accuracy of 98.65% for the ClaMP malware dataset, and 99.86% for the BIG 2015 malware dataset. The results show that, compared to existing malware detection algorithms, the proposed methodologies are more generalizable and detect novel malware samples with a better prediction rate and a lower false positive rate while requiring fewer system resources. When integrated with various features, the proposed static MDS improves the robustness and scalability of malware detection systems. There is still a conflict between malware authors and the anti-malware industry.

Acknowledgment

The authors are grateful to VIT management for providing the Research Seed Grant (AY 2019-20) to execute this work.

The authors would like to express sincere thanks to Mr. J. Kesavardhanan, Founder and Chairman, K7 Computing Pvt., Ltd., and CEO of The Association of Anti-Virus Asia Researchers (AVAR) for providing malware samples from their laboratory for analysis, and for the discussions, helping us out with malware research directions.

References

[1] Malware Statistics & Trends Report |AV-TEST. (2021). Retrieved 27 November 2021, Available from https://www.av-test.org/en/statistics/malware/.

[2] M. Wadkar, F. Di Troia, M. Stamp, 'Detecting malware evolution using support vector machines', *Expert Systems With Applications*, 143, 113022, 2020.

[3] D. Du, Y. Sun, Y. Ma, F. Xiao, 'A novel approach to detect malware variants based on classified behaviors', *IEEE Access*, 7, pp. 81770-81782, 2019.

[4] Y. Ding, X. Xia, S. Chen, Y. Li, 'A malware detection method based on family behavior graph', *Computers & Security*, 73, 73-86, 2018.

[5] N. Azeez, O. Odufuwa, S. Misra, J. Oluranti, R. Damaševičius, 'Windows PE malware detection using ensemble learning', *Informatics*, 8(1), p. 10, 2021.

[6] L. Nataraj, S. Karthikeyan, G. Jacob, B. S. Manjunath, 'Malware images: Visualization and automatic classification', In *Proceedings of the 8th International Symposium on Visualization for Cyber Security*, pp. 1-7, July, 2011.

[7] J. Fu, J. Xue, Y. Wang, Z. Liu, C. Shan, 'Malware visualization for fine-grained classification', *IEEE Access*, 6, 14510-14523, 2018.

[8] R. Quinlan, 'Learning decision tree classifiers', *ACM Computing Surveys*, 28(1), pp. 71–72, 1996.

[9] L. Breiman, 'Random forests', *Machine Learning*, 2001.

[10] P. Geurts, D. Ernst, L. Wehenkel, 'Extremely randomized trees', *Machine Learning*, 63(1), pp. 3–42, 2006.

[11] D. Nielsen, 'Tree boosting with XGBoost-why does XGBoost win "every" machine learning competition?', Master's thesis, NTNU, 2016.

[12] M. Ahmadi *et al.*, 'Novel feature extraction, selection and fusion for effective malware family classification', In *Proceedings of the 6th ACM Conference on Data and Application Security and Privacy*, pp. 183-194, March, 2016.

[13] S. A. Roseline, S. Geetha, S. Kadry, Y. Nam, 'Intelligent vision-based malware detection and classification using deep random forest paradigm', *IEEE Access*, 8, pp.206303-206324, 2020.

[14] J. Hemalatha, S. A. Roseline, S. Geetha, S. Kadry, R. Damasevieius, 'An efficient DenseNet-based deep learning model for malware detection', *Entropy*, 23(3), p. 344, 2021.

[15] S. A. Roseline, A. D. Sasisri, S. Geetha, C. Balasubramanian, 'Towards efficient malware detection and classification using multilayered random forest ensemble technique', In *Proceedings of the 2019 International Carnahan Conference on Security Technology (ICCST)*, IEEE, October, 2019.

[16] S. A. Roseline, G. Hari, S. Geetha, R. Krishnamurthy, 'Vision-based malware detection and classification using lightweight deep learning paradigm', In *Proceedings of the International Conference on Computer Vision and Image Processing*, Springer, Singapore, pp. 62-73, September, 2019.

[17] S. Abijah Roseline, S. Geetha, 'Intelligent malware detection using deep dilated residual networks for cyber security', In *Proceedings of the IGI Global Research Anthology titled Research Anthology on Artificial Intelligence Applications in Security*, IGI Global, pp. 1085-1099, 2021.
[18] A. Kumar, 'ClaMP (classification of malware with PE headers)', Mendeley Data, V1, 2020.

10

Malware and Ransomware Classification, Detection, and Prevention using Artificial Intelligence (AI) Techniques

Md Jobair Hossain Faruk[1,*], Hossain Shahriar[1], Mohammad Masum[1], and Khairul Alom[2]

[1]Kennesaw State University, USA
[2]Northern University Bangladesh, Bangladesh
E-mail: mhossa21@students.kennesaw.edu; hshahria@kennesaw.edu; mmasum@kennesaw.edu; kalom.nub@gmail.com
*Contact Author

Abstract

Rapid technological advancement has made cybersecurity more difficult due to damaging malware and ransomware assaults that pose a critical security danger. When it comes to countering freshly developed, sophisticated, dangerous programs, traditional antimalware and ransomware solutions are highly constrained. In contrast, antimalware and ransomware technology has advanced significantly. There is still a lot of work to bring cutting-edge ideas to fruition. Detecting and blocking malware and ransomware activities is the focus of this study, as is the neural network may be utilized to build new malware solutions. We examine the present malware detection methods, their faults, and how to improve them. A decision tree (DT), random forest (RF), Naive Bayes (NB), logistic regression (LR), and neural network based classifiers were employed to classify ransomware. Using ransomware data, we evaluated our proposed framework for each technique. The experimental results demonstrate that RF classifiers outperform other methods in terms of accuracy (0.99 ± 0.01), F-beta (0.97 ± 0.03), precision scores (0.99 ± 0.00), and NB perform best in recall (0.99 ± 0.00).

Keywords: Malware and ransomware classification, detection and prevention, machine learning, artificial intelligence, software and cyber security.

10.1 Introduction

When it comes to computer viruses, the methods used to develop them have changed as technology has progressed. There is still a lot of worry about malicious programs or attacks like malware and ransomware, which may inflict catastrophic damage to computer systems, data centers, the internet, and mobile apps across various industries and enterprises [1]–[3]. The Apple II machines were the first to be infected with the "Elk Cloner" virus, discovered in 1982 by 15-year-old Rich Skrenta [4]. Since that time, there has been an increase in malware-based applications, which take advantage of bugs in the software. When Elk Cloner and Brain were first created as computer viruses, they were meant to detect flaws rather than damage or destroy machines. When it comes to interrupting computers, stealing data, or even gaining entry into private networks, malware has become increasingly dangerous in recent years [5]. These computer viruses can infect any government agency, data centers, laboratories, commercial, corporate, or organizational programs, or even simply a specified connection. They can spread through any of these methods. Ransomware is a pervasive and complex threat that impacts people throughout the world in various ways. In order to get access to a user's system or data, it locks the screen or encrypts the user's files [2]. [1] Due to the long-term damage caused by uninstalling the ransomware, most ransomware prevents intended victims from accessing computer data by employing an impregnable encrypting mechanism. By refusing the attacker's demands, the data will be lost forever. Ransomware is being repurposed to establish new families, making it more difficult to reverse an infection [7]. Locker ransomware, which prevents access to a computer or device, and crypto-ransomware, which prevents access to files or data, are the two most common types of ransomware. Unless you pay the extortion, you will never be able to recover from these attacks. Traditional methods for detecting ransomware, such as those focused on events, statistics, and data, are woefully underprepared to deal with the problem. As a result, the research community must be protected to the most significant possible degree of safety and security through futuristic technology in the face of such advanced hostile attacks.

There are a variety of technologies and anti-virus programs being developed by security specialists worldwide to help combat malware and ransomware attacks [9]. Still, the exact origins of the first antivirus are a hot topic of discussion. Bernd Robert, a German computer security expert, developed the first antivirus application that successfully eliminated a computer virus in 1987. He came up with a scheme to rid Vienna of him. These viruses infected the files on DOS-based platforms [10]. There are many manual and automatic malware and ransomware detection and prevention solutions available for various platforms, such as mobile devices, servers, gateways, and workstations. The detection process is updated, and prevention begins with proactive measures. Security measures and antivirus must be implemented in the most cutting-edge methods possible, a technological marvel nowadays. Malware and ransomware detection and prevention will be easier by advancing these sectors' research and development efforts. To enhance the detection and prevention of malware and ransomware, AI can enable developers to create rapid, resilient, and scalable malware identification modules. Despite its limitations, artificial intelligence (AI) has the potential to be an effective and potent weapon against even the most complex forms of malware [11]. Malware and ransomware detection, for example, might benefit greatly from machine learning, which is a new area of research [12]. Thanks to machine learning (ML) algorithms, automated detection of malware, including ransomware, is now possible, thanks to machine learning (ML) algorithms [13]. Decision tree (DT), random forest (RF), Naive Bayes, logistic regression, and neural network (NN) based architectures are just a few of the methodologies we use to classify and identify malware and ransomware threats.

Contributions: Experts worldwide are working to create security tools and antivirus software that may be used to avoid, identify, and eradicate viruses, Trojans, worms, malware, and ransomware [9]. As far as we know, the origins of the first antivirus remain unknown. A German computer security expert, Bernd Robert, used an antivirus program to successfully remove a computer virus for the first time in 1987. As part of a strategy to clear Vienna of its inhabitants, .com files were infected by this virus on DOS-based computers. It is possible to use a number of human or automated techniques to identify and prevent malware and ransomware on a wide range of platforms, including mobile devices, servers, gateways, and desktop computers. Technology has made it imperative that new methods for creating effective security measures and anti-viruses be put into place. It will be easier to identify and stop ransomware if these areas are improved. Malware and ransomware detection and prevention can be improved using artificial intelligence (AI) that enables developers to create rapid, robust, and scalable malware identification modules. When used against even the most complex malware, artificial intelligence (AI) has the potential to be an effective and formidable weapon [11]. Malware and ransomware detection, for example, might benefit greatly from machine learning techniques. Malware, including ransomware, may now be detected automatically using ML algorithms based on dynamic characteristics [13]. DT, RF, NB, LR, NN, and LR/NN-based architectures are just a few of the approaches we use to classify and identify malware and ransomware

10.2 Malware And Ransomware

When describing dangerous software or programs confined to computers and the internet, the term "malware" is used. There are many different types of malware, including computer viruses and worms; ransomware; rootkits; Trojans; dialers; adware; spyware; and keyloggers [14]. Malware may also be found in the form of harmful browser helper objects (BHO) used to target computers and web-based applications. When the internet had just four hosts in 1969, viruses began to spread slowly but gradually. According to current estimates, there will be 1.01 billion hosts worldwide in 2019 [15].

Malware is a term used to describe harmful software or programs confined to computers, the internet, and related places. When it comes to malicious programming codes or scripts that are used to attack computers and programs,

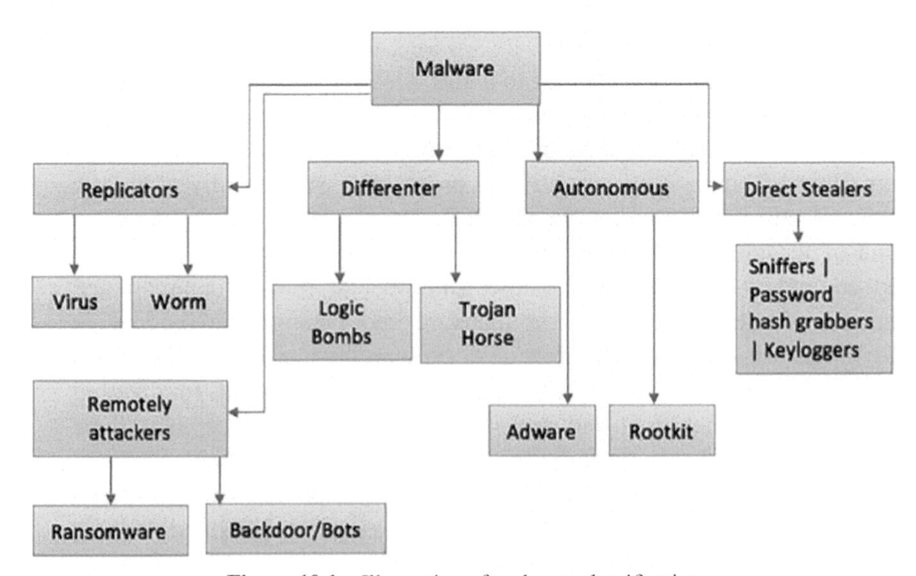

Figure 10.1 Illustration of malware classification

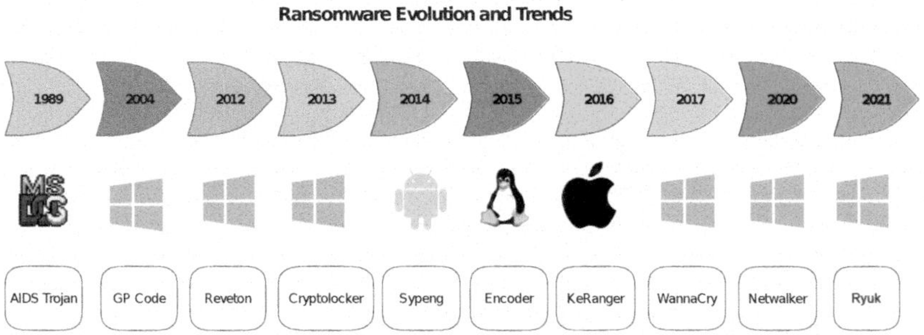

Figure 10.2 Ransomware timeline and trends **[18]**

as well as mobile- or web-based applications in various forms, including ransomware, worms, rootkits, Trojans, and spyware, as well as malicious browser helper objects (BHOs), malware is the term of choice [14]. There were just four hosts on the Internet in 1969, and viruses and the internet's rise started slowly but gradually. It is predicted that there will be 1.01 billion hosts [15] in use by the year 2019.

10.3 Artificial Intelligence

All organizations are eager to reap the benefits of artificial intelligence (AI) because of its ability to do tasks formerly reserved for the human mind, reducing costs and increasing efficiency [19]. One definition of artificial intelligence (AI) is the rapidly evolving development of computer systems capable of performing activities that can only be performed by human intellect. Rather than becoming a replacement for human intellect, academics believe that AI may be utilized to enhance human intelligence (IA). This gives AI strategic importance as a potential primary driver of the current technological revolution. Because of this, artificial intelligence (AI) could be used to produce projects based on cognitive processes such as augmentation, conception, conscious research, and analysis of passion-inspiring information [21], as well as creative problem-solving in sectors such as Big Data, security, and business analytics. AI might detect malware and ransomware categorization, detection, and prevention. Consequently, artificial intelligence (AI) may be created by various methods, including machine learning [22]. Figure 10.3 depicts two instances of artificial intelligence and its applications: machine learning and natural language processing. In the subject of machine learning (MI), which uses logical or binary processes to learn a task from a series

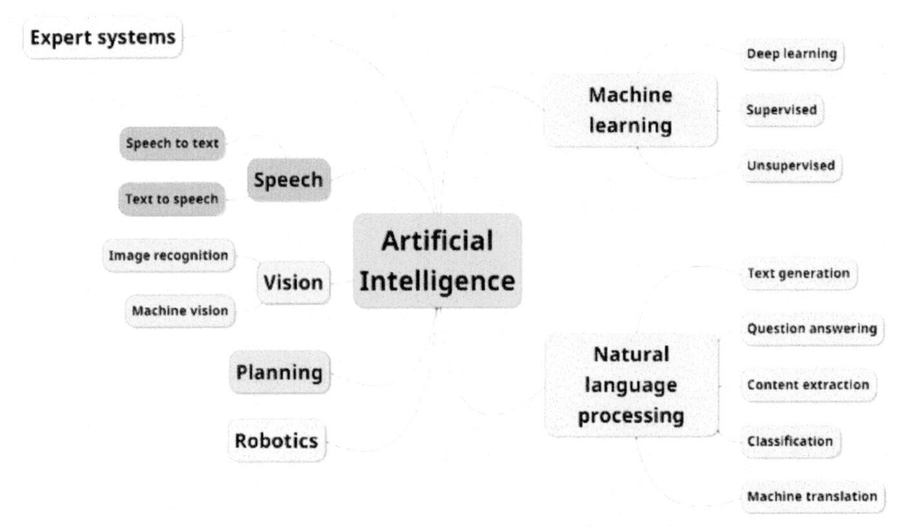

Figure 10.3 Types and uses of artificial intelligence

of examples without being explicitly programmed, computers can achieve artificial intelligence (AI) without having to be explicitly programmed [23]. Computer viruses and malware may be prevented using artificial intelligence (AI) by assessing and learning from situations utilizing situational methods for machine learning [24].

10.4 Related Work

For malware and ransomware detection, the most important purpose is to defend the system against damaging assaults. Malware and ransomware may now be detected using a variety of methods. As malware technology advances, artificial intelligence becomes increasingly important for effective and reliable malware protection systems. Finding the malicious source code is the first step in identifying malware. In this section, we present some relevant research on malware and ransomware.

The world's largest malware source code database, SourceFinder, was developed by researchers Rokon *et al.* [25] to find malware source code repositories. SourceFinder was shown to have an accuracy rate of 89% and a recall rate of 86% when used to identify 7504 malware source code repositories before assessing their features and attributes. Machine learning techniques often assist malware detection. It is aided by the author's machine learning and data mining approaches. Others, for example, give a comprehensive study of static, dynamic, and hybrid methodologies and evaluation of malware detection methods, such as Sharma *et al.* [26]. In addition, the research studies and assesses a wide range of data mining and machine-learning-based malware detection solutions. On the other hand, Sharma *et al.* [27] advocate using machine learning to detect malware by looking for instances of the opcode. Using the Kaggle Microsoft malware classification challenge dataset, the researchers additionally examined five classifiers, including LMT, REPTree, random forest, NBT, and J48Graft. The demonstration shows that the proposed method can correctly identify malware with 99% accuracy. Machine learning approaches, notwithstanding the difficulties of applying them to intrusion detection, such as atypical computing paradigms and unconventional evasion tactics, are described by Saad *et al.* as having three main drawbacks that restrict the efficacy of malware detectors. Additionally, the researchers address how to overcome the limitations of behavioral analysis in next-generation antimalware systems. Using artificial intelligence, Syam and Vankata [29] developed a virtual analyst to guard against threats and gather the necessary steps in a separate inquiry. It improves the algorithm over time, making it more powerful and efficient. In order to automatically update the system, the researchers used analyst feedback to turn unsupervised data into supervised data and make a distinction between the two. Cloud computing, network detection systems, virtual machines, and hybrid approaches and technologies are all employed in malware detection in addition to machine learning. The use of deep learning and artificial intelligence to detect malware is becoming more common. Baptista *et al.* [30] have developed a new method for detecting malware that combines binary visualization with self-organizing incremental neural networks. Experimental findings show that ransomware detection accuracy was 91.7% and 94.1% for a variety of file formats that included Portable Document File (.pdf) and Microsoft Document File (.doc). When used in conjunction with an incremental detection rate, the authors' method worked well enough to allow for the quick and accurate identification of new malware. Mohammad Masum and his colleagues at Kennesaw State University [31]. Ransomware has been classified using traditional detection methods. Analysis of distinct ransomware families may be done using a well-defined behavioral framework, and most of these families share behavioral traits such as payload persistence, stealth tactics, and network activity. Abiola and Marhusin proposed a signature-based detection methodology for malware by extracting the Brontok worms and utilizing an *n*-gram technique to break down the signatures [32]. Thanks to the design, malware can be identified and a safe solution may be created. The authors in [33] proposed a static and dynamic design, also known as a behavior-based architecture, to get around this limitation. The application's code is analyzed statically for potentially dangerous behaviors. For instance, processes monitored using dynamics-based analysis can be flagged as having malicious intent and shut down if any suspicious behavior is detected. Static and dynamic analyses have limitations in finding undiscovered malware and ineffectiveness against obfuscation, large variance output, and targeted assaults. Using a huge amount of unlabeled data and a limited amount of labeled data, Noorbehbahani and Saberi [13] focused on semi-supervised learning to identify ransomware. Ransomware in the CICAndMal 2017 dataset was studied using many feature selection and semi-supervised classification algorithms. While other semi-supervised classification approaches for ransomware detection are better than random forest, they are not as effective. The evaluation of alternative feature selection and semi-supervised classification algorithms for ransomware analysis was the goal of CICAndMal 2017. The semi-supervised classification of ransomware that uses the random forest as a foundation classifier outperforms the other semi-supervised classification methods. Machine learning ideas must be used in ransomware detection and prevention in order to enhance current approaches. The Argus server and client apps were integrated into a network intrusion detection system called Biflow, which uses a unique flow-oriented method known as Biflow to identify ransomware. In order to categorize the datasets, six feature selection strategies were employed, and supervised machine learning was applied to increase the detection module's accuracy and performance. With the detection of malware and ransomware, random forest is one of the most frequently utilized machine learning algorithms, and for a good reason. DNAact-Ran, a digital DNA sequencing engine, was used by Khan *et al.* [35] to propose a ransomware detection approach based on sequencing design restrictions and the k-mer frequency vector. The framework was tested on 582 instances of DNAact-Run ransomware and 942 instances of goodware to determine its precision, recall, f-measure, and accuracy. A technique created by Poudyalwe *et al.* [36] employs machine learning to grasp the function of malware code portions better. The algorithm detected ransomware in 76% to 97% of cases. Ganta *et al.* [37] proposed an anti-ransomware detection system based on machine learning. Various classification approaches, including ex-random forest, decision tree, logistic regression, and the KNN algorithm, were employed to find ransomware concealed in executable files. EldeRan, a machine-learning-based technique for dynamically evaluating and categorizing ransomware, was suggested by Sgandurra *et al.* [38]. Features selection and classification are both done in the Cuckoo Sandbox environment by EldeRan, which utilizes two machine learning components: long-short term memory (LSTM) networks, developed by Maniath *et al.* [39], were used to classify ransomware using a binary sequence API call categorization technique. Windows API calls, Registry Key Actions, File System Actions, the set of file operations done by file extension, Directory Operations, Dropped Files, and Strings are used to retrieve and dynamically analyze datasets. In total, 582 ransomware files from 11 distinct families and 942 goodware applications were used in the system's testing, which resulted in an accuracy rate of 94.95%. A dynamic analysis technique gathered API calls from the altered log in a sandbox environment. Analyzing many malware datasets shows that the LSTM-based approach suggested has an accuracy rate of 96.67% in automatically categorizing ransomware activities. Strengthening the LSTM network can further increase the overall accuracy. Machine learning may identify new ransomware variations and families with such precision. It is possible to create a revolutionary dynamic detection method for identifying ransomware using a deep neural network (DNN). Deep neural network-based intrusion detection may benefit from using a novel Bayesian optimization-based framework for automatic optimization of hyperparameters presented by the researchers [40]. Researchers have suggested a dynamic malware detection system based on deep neural network (DNN) and convolutional neural network (CNN) techniques. The machine learning model is built using long short-term memory (LSTM). A novel technique identified malware samples using CNNs and the LSTM network. According to the evaluation report, the combination of DNN and LSTM accurately identifies new malware with 91.63% accuracy. Android malware may also be detected using deep learning. The deep learning framework (mechanized as a deep learner) for Android malware classification created by Masum and Shahriar, which surpasses cutting-edge machine learning techniques, is Droid-NNet. The Droid-NNet suggests powerful and effective malware detection on the Android platform based on an evaluation report on two datasets of Android apps: Malgenome-215 and Drebin-215 [42].

10.5 Malware Detection Using AI

Modern systems face an increasing danger from malware, which continues to evolve and diversify, demanding the development of unique solutions to protect against cyber-criminals. To add, AI is a dynamic field that is always growing while also delivering impressive outcomes across a wide range of application fields.

Antimalware systems that can overcome the limits of current preventative technologies will benefit from AI advancements. Malware detection systems have been developed that monitor both harmful and non-malicious software to analyze it properly.

Antimalware systems that can overcome the limits of current preventative technologies will benefit from AI advancements. Malware detection systems have been developed to monitor both harmful and non-malicious software to analyze it properly.

Antimalware systems that can overcome the limits of current preventative methods will benefit from improvements in artificial intelligence (AI). In order to properly analyze harmful and non-malicious software, researchers developed malware detection algorithms that monitor both.

- It is possible to identify and detect assaults by looking for specific patterns using a signature-based detection approach, which comprises four components [43]. When a virus is discovered in a database, it is

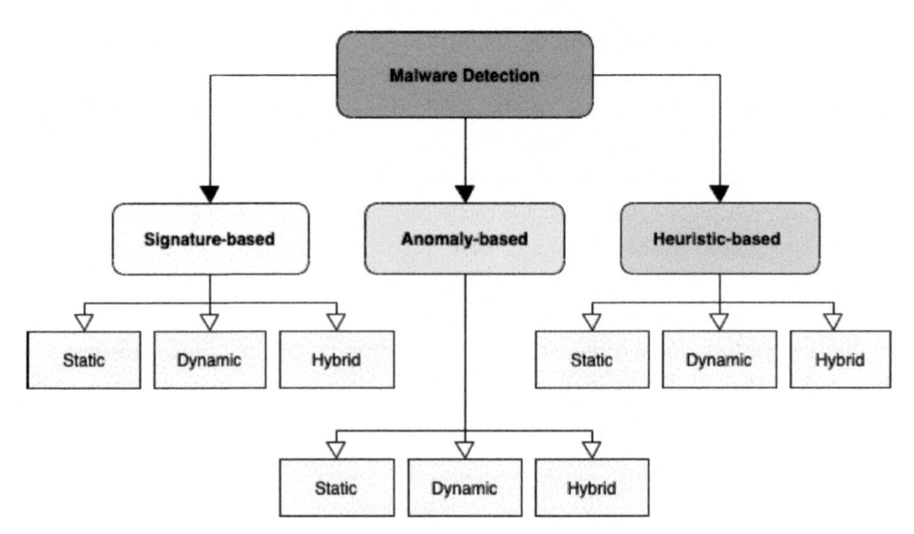

Figure 10.4 Classification of malware detection techniques

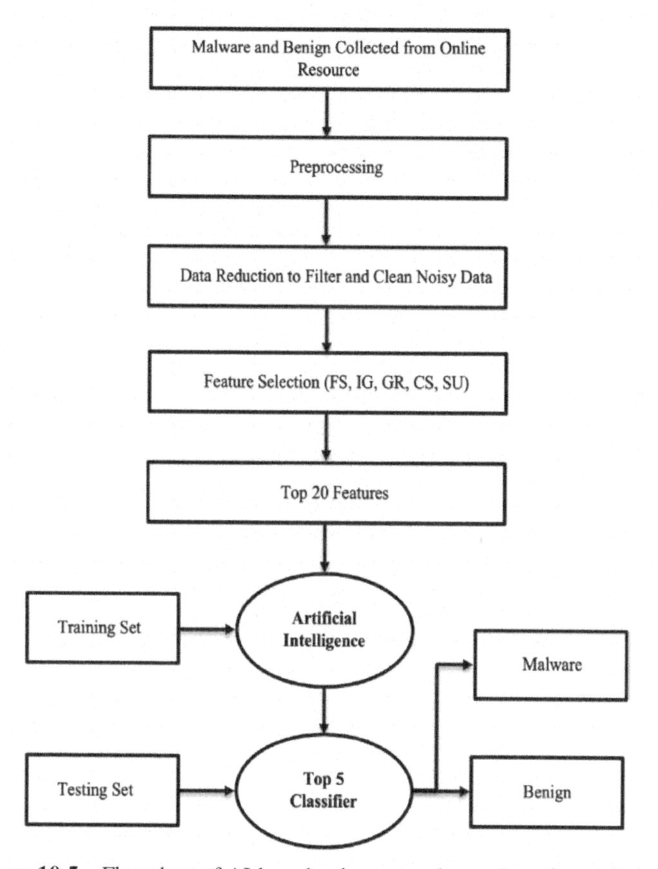

Figure 10.5 Flowchart of AI-based unknown malware detection techniques

found by scanning the file and matching its contents to a database of viral signatures. The file is infected if the data matches the database. This technique has a major benefit in identifying known malware, but it has certain disadvantages in detecting new malware [44]. Using a statistical traffic model, the IDS (intrusion detection system) keeps a database. When an IDS detects suspicious activity, it compares the traffic from various sources to a predetermined set of traffic patterns and notifies an administrator of its results.

- The detection of network intrusions based on anomalies: Unwanted network activity cannot be tolerated with anomaly-based network intrusion detection [45]. By applying classification algorithms to a system's behaviors, anomaly-based approaches go beyond the restrictions of signature-based approaches and allow users to identify any type of malware, known or undiscovered. Moving from pattern-based to classification-based detection of normal and aberrant activity enhances the likelihood of catching malware. Almost all anomaly-based network intrusion detection systems make use of the operational phases (ANIDS). There are known attack signatures in a database, and the common signatures come from the numerous packets that connect to the database. If an unknown signature matches one of the known signatures, the system administrator is notified.

- To enhance the effectiveness of signature and anomaly-based detection systems, heuristic-based detection techniques use artificial intelligence (AI). A machine learning method known as the genetic algorithm was integrated with a neural network to boost the malware detection system's categorization process and respond to environmental changes. Inheritance, selection, and combination are methods used to find the best possible answers from various angles, all without having any prior understanding of the system. The heuristic strategy beats prior methods because it incorporates statistical and mathematical ideas.

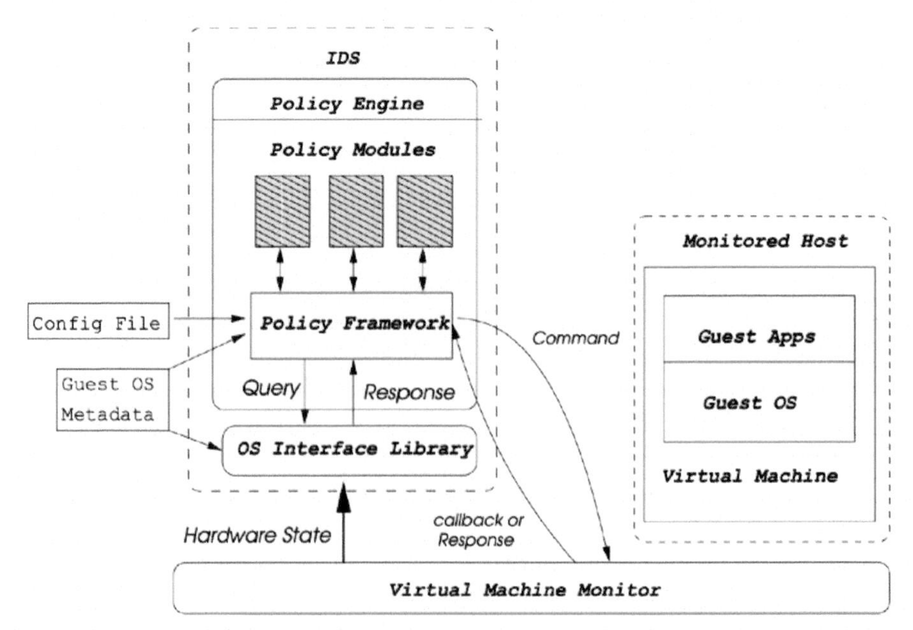

Figure 10.6 A high-level view of the VMI-base

- Adopting AI to detect malware: Garfinkel and Rosenblum [46] proposed a virtual machine monitoring approach for detecting malicious software. The assessment shows that it is possible to get a specific restriction on contacts between the host and the effective program using a virtual machine monitor. While keeping the host-based intrusion detection system (IDS) transparent, the IDS is kept away from the host for improved assault resistance as seen in Figure 10.6 [46].

However, the suggested method is constrained by the possibility of errors and tamper resistance. In contrast, Li *et al.* [47] provide a malware classification based on a graph convolutional network that can adapt to various pathogen properties. The technique pulls the API call sequence from the malware's code based on a directed cycle graph. It generates a convolutional network using the Markov chain and principal component analysis approach to design a classifier. An evaluation and comparison of its performance are also part of the method's function. GCN acts as the base for the malware detection system framework. It was found that the highest accuracy was 98.32%, which is better than other current strategies in terms of FPR and accuracy.

10.6 Ransomware Detection

10.6.1 Methodology

When identifying ransomware, we relied on well-known machine learning classifiers such as decision trees, random forests, Naive Bayes, logistic regression, and neural network design. Figure 10.7 shows the model's structure. To make the model more general, we used a 10-fold cross-validation strategy. Different scale factors in the ransomware data were standardized to provide one common measurement range. In order to distinguish ransomware from real observations, a feature selection approach was utilized to extract multiple key properties from the data. As a final step, we analyzed the models' accuracy, F-beta score, and other measures such as precision, recall, and the area under the ROC curve to see how well they performed.

10.6.2 Experiments and Result

- In the dataset [49], there are 138,047 samples with 54 characteristics, with 70% of the samples being malware and 30% legal observations. The dataset distribution is shown in Figure 10.8 (left).

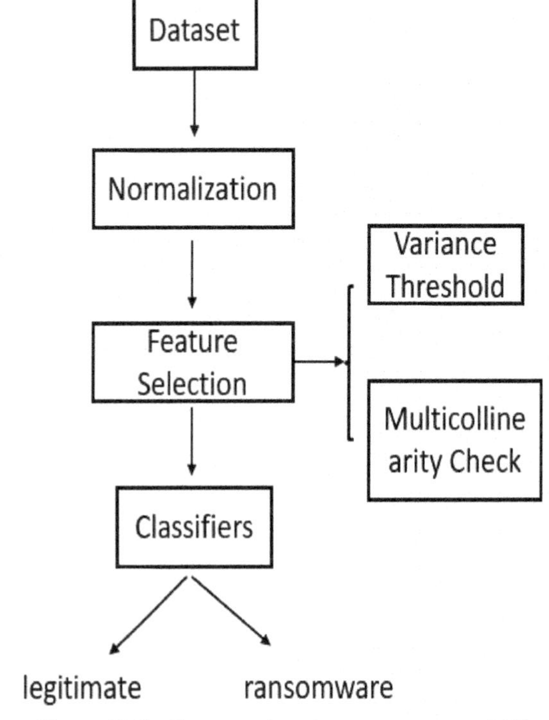

Figure 10.7 Framework to detect ransomware **[48]**

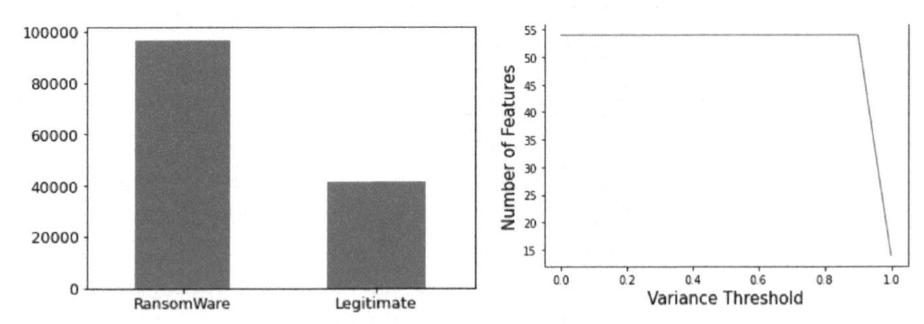

Figure 10.8 Distribution of the dataset (left) and number of features with varying variance thresholds (right)

- **Feature selection:** Standardization of the Z-score method resulted in each variable being placed on a scale with a standard deviation of 1. Variance thresholds and inflation factors were used to exclude low-variability and strongly related variables from the dataset. A variance threshold of 1 was created after removing low variation features from the dataset. The number of characteristics went from 54 to 13 when the threshold was set to 1. It is seen in Figure 10.8 how many different features have differing thresholds for variation (right).

We used the variance inflation factor to test for multicollinearity in the high variance features in the second round of feature selection (VIF). If the VIF score exceeds 10, a feature is found.

Features: Mean and maximum raster widths, which have VIF scores of 19.52 and 198, respectively, indicate multicollinearity. One of these variables was selected at random. All of the characteristics in Table 10.1 have VIF scores below the threshold for high variation. Finally, the classifiers are used to identify ransomware based on the previously mentioned 12 features.

- **Evaluation metrics:**

Recall: The proportion of correct positive predictions across all positive samples. Mathematically:

$$\text{Recall} = \frac{\text{TP}}{\text{TP} + \text{FN}} \tag{10.1}$$

In this example, TP stands for true positive (number of correct positive predictions) and FN stands for false negative (number of incorrect negative predictions) (quantity of misclassified positive predictions).

Prediction accuracy: The percentage of predicted positives that are identified. Mathematically:

$$\text{Precision} = \frac{\text{TP}}{\text{TP} + \text{FP}} \tag{10.2}$$

F_1 score: The harmonic means of precision and recall. F_1score is better than the accuracy metric for imbalanced data.

$$\text{F}_1 = 2 \times \frac{\text{Precision} \times \text{Recall}}{\text{Precision} + \text{Recall}} \tag{10.3}$$

The *F*-beta score is the weighted harmonic mean of precision of recall where *F*-beta value at 1 means perfect score (perfect precision and recall) and 0 is worst.

$$\text{F}_\beta = (1 + \beta^2)\frac{\text{Precision} \times \text{Recall}}{(\beta^2 \times \text{Precision}) + \text{Recall}} \tag{10.4}$$

When $B = 1$, F-beta is F_1 score, the B parameter determines the weight of precision and recall. $B < 1$ can be picked, if we want to give precision, while $B > 1$ values give more weight to recall.

- **Experimental settings:** We compared our model against the LR, NB, RF, and DT approaches to see how well it performed. Authentic and malware samples were evenly distributed throughout the two datasets' training and testing portions. Data from training and testing was used to develop and evaluate each of the models put to the test. A 10-fold cross-validation was done on each model to ensure uniformity. Both datasets were compared using the RF, LR, NB, DT, and NN-based classifiers. Scikit-learn and hyperparameter settings were used to create the algorithms and then coded in Python. It has four levels: an input layer, two hidden layers, and an output layer. The neural-network-based architecture. Because this is a binary classification issue, we employed the ReLu activation function in the hidden layers and the sigmoid function in the output layer. The optimizer and loss function were given "Adam" and "binary cross-entropy," respectively. Early stopping techniques were used to halt training when performance on test data deteriorated. For early halting, we set the minimal delta to 1e-3 and the patience to 5 to monitor validation loss. This checks for even the tiniest changes in the monitored amount (the number of epochs that produced the monitored quantity with no improvement after which training will be stopped). This was the starting pace of learning.
- **Results:** For ransomware datasets (NN), decision trees (DT), random forests (RF), Naive Bayes, logistic regressions (LR), and neural networks (NN) were used to analyze the ransomware datasets. The performance of the models was evaluated using the F-beta score. Each experiment was cross-validated for a total of ten times. The identical setup was used for all datasets to maintain uniformity. Table 10.1 shows the accuracy, F-beta score, recall, and precision of the models. When it comes to precision, accuracy, and F-beta scores, random forest surpasses other models by a wide margin. Although having the highest recall (0.990.00), the NB classifier performed poorly on other metrics.

The DT and NN classifiers outperform the RF one in terms of performance. In spite of having an appropriate accuracy score when compared to other algorithms, LR's F-beta and recall ratings are not good enough. Tenfold and mean ROC curves for each classifier are shown in Figures 10.9 and 10.10. The RF, LR, and NN had the highest mean AUC values of 0.99, while the NB had the lowest AUC score (mean AUC of 0.73).

Table 10.1 Experimental results analysis of different classifiers

Classifiers	Accuracy	*F*-beta	Recall	Precision
DT	0.98 ± 0.01	0.94 ± 0.05	0.94 ± 0.05	0.98 ± 0.00
RF	$\mathbf{0.99 \pm 0.01}$	$\mathbf{0.97 \pm 0.03}$	0.97 ± 0.03	$\mathbf{0.99 \pm 0.00}$
NB	0.35 ± 0.03	0.97 ± 0.03	$\mathbf{0.99 \pm 0.00}$	0.31 ± 0.01
LR	0.96 ± 0.02	0.89 ± 0.07	0.89 ± 0.07	0.96 ± 0.00
NN	0.97 ± 0.01	0.95 ± 0.05	0.95 ± 0.05	0.97 ± 0.00

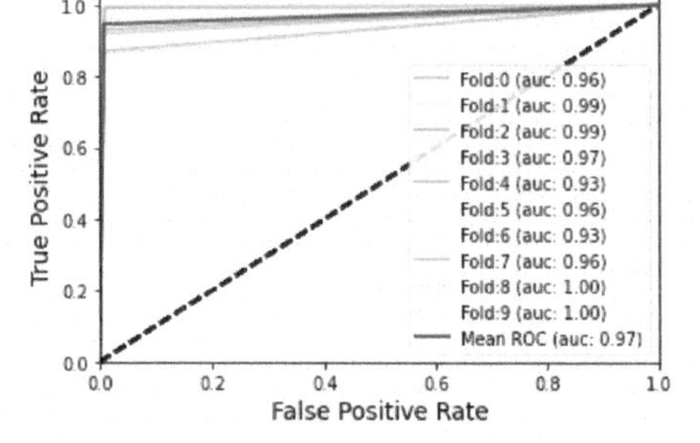

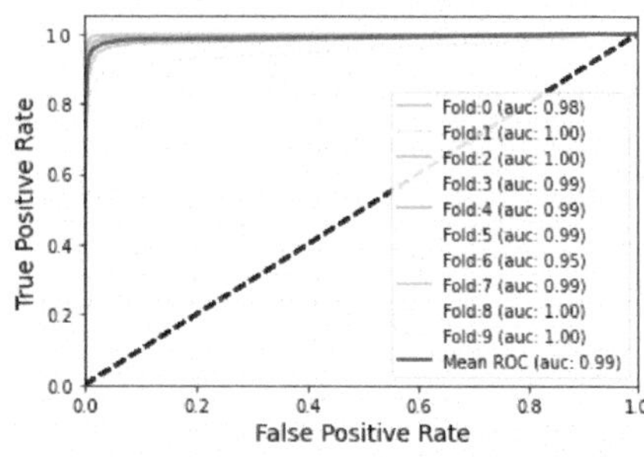

Figure 10.9 ROC curve for decision tree classifier (left) and random forest classifier (right)

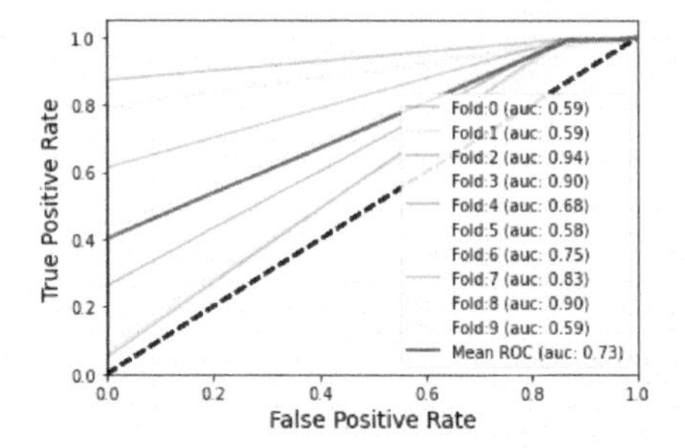

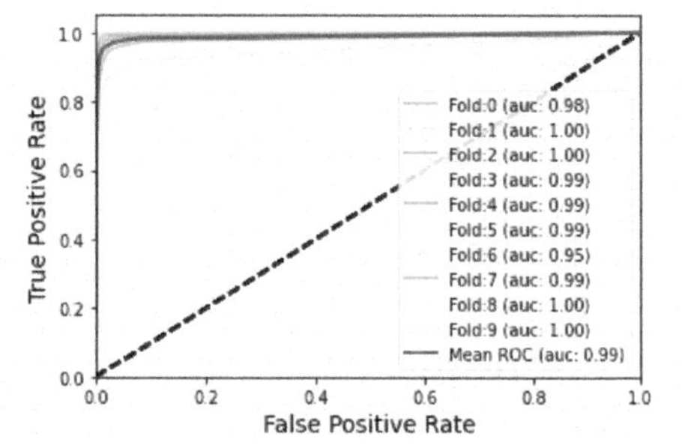

Figure 10.10 Naïve Bayes classifier (left) and neural network classifier

10.7 Conclusion

Financial institutions, corporations, and individuals are becoming increasingly vulnerable to malware and ransomware. Malware and ransomware attacks can be minimized by implementing automated detection, classification, and prevention systems. We use a number of machine learning algorithms, including classifiers based on neural networks, to create a unique framework for the successful categorization and detection of ransomware. We tried out several classifiers such as DDT, RF, NB, LR, and NN with a ransomware dataset. Tenfold cross-validation showed that random forest beat other classifiers in accuracy, *F*-beta, and precision while maintaining a reasonable consistency level. Antimalware and anti-ransomware systems, which can identify and prevent malware and ransomware assaults and security flaws in software applications, will also benefit from AI's ability to recognize and prevent these issues. Malware, ransomware, AI, and narrative were all covered in detail in this chapter.

Acknowledgment

This work is partially supported by the U.S. National Science Foundation Awards #2100115 and #1723578, SunTrust Fellow Award#ST22-01. Any opinions, findings, and conclusions or recommendations expressed in this material are those of the authors and do not necessarily reflect the views of the National Science Foundation or SunTrust.

References

[1] K. Shaukat, S. Luo, V. Varadharajan, I. A. Hameed, and M. Xu, "A Survey on Machine Learning Techniques for Cyber Security in the Last Decade," IEEE Access, vol. 8, pp. 222310–222354, 2020, doi: 10.1109/ACCESS.2020.3041951.

[2] N. Shah and M. Farik, "Ransomware-Threats, Vulnerabilities and Recommendations," Int. J. Sci. Technol. Res., 2017. [Online]. Available: https://www.ijstr.org/final-print/june2017/Ransomware-Threats-Vulnerabilities-And-Recommendations.pdf.

[3] M. J. Hossain Faruk et al., "Malware Detection and Prevention using Artificial Intelligence Techniques," Proc. 2021 IEEE Int. Conf. Big Data, Big Data, 2021, [Online]. Available: https://www.researchgate.net/publication/357163392.

[4] S. Levy and J. R. Crandall, "The Program with a Personality: Analysis of Elk Cloner, the First Personal Computer Virus," 2020. [Online]. Available: http://arxiv.org/abs/2007.15759.

[5] I. Khan, "An Introduction to Computer Viruses: Problems and Solutions," Libr. Hi Tech News, vol. 29, no. 7, pp. 8–12, 2012, doi: 10.1108/07419051211280036.

[6] F. Noorbehbahani, F. Rasouli, and M. Saberi, "Analysis of Machine Learning Techniques for Ransomware Detection," Proc. 16th Int. ISC Conf. Inf. Secur. Cryptology, Isc., pp. 128–133, 2019, doi: 10.1109/ISCISC48546.2019.8985139.

[7] U. Adamu and I. Awan, "Ransomware Prediction using Supervised Learning Algorithms," Proc. 2019 Int. Conf. Futur. Internet Things Cloud, FiCloud 2019, pp. 57–63, 2019, doi: 10.1109/FiCloud.2019.00016.

[8] K. Savage, P. Coogan, and H. Lau, "The Evolution of Ransomware," Res. Manag., vol. 54, no. 5, pp. 59–63, 2015. [Online]. Available: http://openurl.ingenta.com/content/xref?genre=article&issn=0895-6308&volume=54&issue=5&spage=59.

[9] M. Joshi, B. V Patil, and M. J. Joshi, "A Study of Past, Present Computer Virus & Performance of Selected Security Tools," 2012. [Online]. Available: https://www.researchgate.net/publication/288725457.

[10] "History of the {Antivirus}," Hotspot Shield VPN, 2017. [Online]. Available: https://www.hotspotshield.com/blog/history-of-the-antivirus/.

[11] G. Vigna, "How AI Will Help in the Fight Against Malware," TechBeacon, 2017. [Online]. Available: https://techbeacon.com/how-ai-will-help-fight-against-malware.

[12] D. W. Fernando, N. Komninos, and T. Chen, "A Study on the Evolution of Ransomware Detection using Machine Learning and Deep Learning Techniques," IoT, vol. 1, no. 2, pp. 551–604, 2020, doi: 10.3390/iot1020030.

[13] F. Noorbehbahani and M. Saberi, "Ransomware Detection with Semi-Supervised Learning," Proc. 2020 10th Int. Conf. Comput. Knowl. Eng. ICCKE, pp. 24–29, 2020, doi: 10.1109/ICCKE50421.2020.9303689.

[14] M. Ahmad, "Malware in Computer Systems: Problems and Solutions," Int. J. Informatics Dev., vol. 9, p. 1, 2020, doi: 10.14421/ijid.2020.09101.

[15] S. Gupta, "Types of Malware and its Analysis," Int. J. Sci. Eng. Res., vol. 4, no. 1, pp. 1–13, 2013.

[16] S. Poudyal, Z. Akhtar, D. Dasgupta, and K. D. Gupta, "Malware Analytics: Review of Data Mining, Machine Learning and Big Data Perspectives," Proc. 2019 IEEE Symp. Ser. Comput. Intell. SSCI, pp. 649–656, 2019, doi: 10.1109/SSCI44817.2019.9002996.

[17] FBI's IC3, "2020 Internet Crime Report," Fed. Bur. Investig. Internet Crime Complain. Cent., pp. 1–28, 2020. [Online]. Available: https://www.ic3.gov/Media/PDF/AnnualReport/2020_IC3Report.pdf.

[18] A. Kapoor, A. Gupta, R. Gupta, S. Tanwar, G. Sharma, and I. E. Davidson, "Ransomware Detection, Avoidance, and Mitigation Scheme:

A Review and Future Directions," Sustainability, vol. 14, no. 1, 2022, doi: 10.3390/su14010008.

[19] H. Hassani, E. S. Silva, S. Unger, M. TajMazinani, and S. Mac Feely, "Artificial Intelligence (AI) or Intelligence Augmentation (IA): What Is the Future?," AI, vol. 1, no. 2, pp. 143–155, 2020, doi: 10.3390/ai1020 008.

[20] A. Irizarry-Nones, A. Palepu, and M. Wallace, "Artificial Intelligence (AI)," CISSE, 2019. [Online]. Available: https://cisse.info/pdf/2019/rr _01_artificial_intelligence.pdf.

[21] S. Ahn et al., "A Fuzzy Logic Based Machine Learning Tool for Supporting Big Data Business Analytics in Complex Artificial Intelligence Environments," Proc. IEEE Int. Conf. Fuzzy Syst., vol. 2019, June, 2019, doi: 10.1109/FUZZ-IEEE.2019.8858791.

[22] J. Alzubi, A. Nayyar, and A. Kumar, "Machine Learning from Theory to Algorithms: An Overview," J. Phys. Conf. Ser., vol. 1142, no. 1, 2018, doi: 10.1088/1742-6596/1142/1/012012.

[23] T. Oladipupo, "Machine Learning Overview," New Adv. Mach. Learn., 2010, doi: 10.5772/9374.

[24] A. Cranage, "Getting Smart About Artificial Intelligence," Wellcome Sanger Inst., 2019. [Online]. Available: https://sangerinstitute.blog/201 9/03/04/getting-smart-about-artificial-intelligence.

[25] M. O. F. Rokon, R. Islam, A. Darki, E. E. Papalexakis, and M. Faloutsos, "SourceFinder: Finding Malware Source-Code from Publicly Available Repositories in GitHub," Proc. 23rd Int. Symp. Res. Attacks, Intrusions Defenses RAID 2020, pp. 149–163, 2020.

[26] N. Sharma and B. Arora, "Data Mining and Machine Learning Techniques for Malware Detection," Adv. Intell. Syst. Comput., vol. 1187, pp. 557–567, 2021, doi: 10.1007/978-981-15-6014-9_66.

[27] S. Sharma, C. Rama Krishna, and S. K. Sahay, "Detection of Advanced Malware by Machine Learning Techniques," Adv. Intell. Syst. Comput., vol. 742, pp. 333–342, 2019, doi: 10.1007/978-981-13-0589-4_31.

[28] S. Saad, W. Briguglio, and H. Elmiligi, "The Curious Case of Machine Learning in Malware Detection," Proc. 5th Int. Conf. Inf. Syst. Secur. Priv. ICISSP 2019, pp. 528–535, 2019, doi: 10.5220/00074707052805 35.

[29] S. A. Repalle and V. R. Kolluru, "Intrusion Detection System using AI and Machine Learning Algorithm," Int. Res. J. Eng. Technol., vol. PAPER 2, no. 12, pp. 1709–1715, 2017.

[30] I. Baptista, S. Shiaeles, and N. Kolokotronis, "A Novel Malware Detection System based on Machine Learning and Binary Visualization," Proc. 2019 IEEE Int. Conf. Commun. Work. ICC Work. 2019, 2019, doi: 10.1109/ICCW.2019.8757060.

[31] M. Masum et al., "Bayesian Hyperparameter Optimization for Deep Neural Network-Based Network Intrusion Detection," Proc. 2021 IEEE Int. Conf. Big Data (Big Data), pp. 5413–5419, 2022, doi: 10.1109/bigdata52589.2021.9671576.

[32] A. M. Abiola and M. F. Marhusin, "Signature-Based Malware Detection using Sequences of N-Grams," Int. J. Eng. Technol., vol. 7, no. 4, pp. 120–125, 2018, doi: 10.14419/ijet.v7i4.15.21432.

[33] D. Nieuwenhuizen, "A Behavioural-Based Approach to Ransomware Detection," MWR Labs, 2017. [Online]. Available: https://labs.f-secure.com/assets/resourceFiles/mwri-behavioural-ransomware-detection-2017-04-5.pdf.

[34] Y. L. Wan, J. C. Chang, R. J. Chen, and S. J. Wang, "Feature-Selection-Based Ransomware Detection with Machine Learning of Data Analysis," Proc. 2018 3rd Int. Conf. Comput. Commun. Syst. ICCCS 2018, pp. 392–396, 2018, doi: 10.1109/CCOMS.2018.8463300.

[35] F. Khan, C. Ncube, L. K. Ramasamy, S. Kadry, and Y. Nam, "A Digital DNA Sequencing Engine for Ransomware Detection using Machine Learning," IEEE Access, vol. 8, pp. 119710–119719, 2020, doi: 10.1109/ACCESS.2020.3003785.

[36] S. Poudyal, K. P. Subedi, and D. Dasgupta, "A Framework for Analyzing Ransomware using Machine Learning," Proc. 2018 IEEE Symp. Ser. Comput. Intell. SSCI 2018, pp. 1692–1699, 2019, doi: 10.1109/SSCI.2018.8628743.

[37] V. G. Ganta, G. Venkata Harish, V. Prem Kumar, and G. Rama Koteswar Rao, "Ransomware Detection in Executable Files using Machine Learning," Proc. 5th IEEE Int. Conf. Recent Trends Electron. Inf. Commun. Technol. RTEICT 2020, pp. 282–286, 2020, doi: 10.1109/RTEICT49044.2020.9315672.

[38] D. Sgandurra, L. Munoz-González, R. Mohsen, and E. C. Lupu, "Automated Dynamic Analysis of Ransomware: Benefits, Limitations and use for Detection," 2016. [Online]. Available: http://arxiv.org/abs/1609.03020.

[39] S. Maniath, A. Ashok, P. Poornachandran, V. G. Sujadevi, A. U. P. Sankar, and S. Jan, "Deep Learning LSTM based Ransomware

Detection," Proc. 2017 Recent Dev. Control. Autom. Power Eng. RDCAPE 2017, pp. 442–446, 2018, doi: 10.1109/RDCAPE.2017.8358312.

[40] M. Masum et al., "Bayesian Hyperparameter Optimization for Deep Neural Network-Based Network Intrusion Detection," Proc. 2021 IEEE Int. Conf. Big Data, Big Data, 2021, [Online]. Available: https://www.researchgate.net/publication/357164131.

[41] H. Ghanei, F. Manavi, and A. Hamzeh, "A Novel Method for Malware Detection Based on Hardware Events using Deep Neural Networks," J. Comput. Virol. Hacking Tech., vol. 17, no. 4, pp. 319–331, 2021, doi: 10.1007/s11416-021-00386-y.

[42] M. Masum and H. Shahriar, "Droid-NNet: Deep Learning Neural Network for Android Malware Detection," Proc. 2019 IEEE Int. Conf. Big Data, Big Data, pp. 5789–5793, 2019, doi: 10.1109/BigData47090.2019.9006053.

[43] O. O. Cyril, T. Elmissaoui, M. C. Okoronkwo, M. Ihedioha Uchechi, C. H. Ugwuishiwu, and O. B. Onyebuchi, "Signature Based Network Intrusion Detection System using Feature Selection on Android," Int. J. Adv. Comput. Sci. Appl., vol. 11, no. 6, pp. 551–558, 2020, doi: 10.14569/IJACSA.2020.0110667.

[44] Y. Ye, T. Li, Q. Jiang, Z. Han, and L. Wan, "Intelligent File Scoring System for Malware Detection from the Gray List," Proc. ACM SIGKDD Int. Conf. Knowl. Discov. Data Min., pp. 1385–1393, 2009, doi: 10.1145/1557019.1557167.

[45] V. Jyothsna and K. Munivara Prasad, "Anomaly-Based Intrusion Detection System," Comput. Netw. Secur., 2020, doi: 10.5772/intechopen.82287.

[46] T. Garfinkel and M. Rosenblum, "A Virtual Machine Introspection Based Architecture for Intrusion Detection," Proc. Netw. Distrib. Syst. Secur., vol. 1, pp. 253–285, 2003. [Online]. Available: http://citeseerx.ist.psu.edu/viewdoc/download?doi=10.1.1.11.8367&rep=rep1&type=pdf%5Cnhttp://www.isoc.org/isoc/conferences/ndss/03/proceedings/papers/13.pdf.

[47] S. Li, Q. Zhou, R. Zhou, and Q. Lv, "Intelligent Malware Detection Based on Graph Convolutional Network," J. Supercomput., 2021, doi: 10.1007/s11227-021-04020-y.

[48] M. Masum, M. J. Hossain Faruk, and H. Shahriar, "Ransomware Classification and Detection With Machine Learning Algorithms," Proc.

IEEE Annu. Comput. Commun. Work. Conf., Jan. 2022. [Online]. Available: https://www.researchgate.net/publication/357335797.

[49] M. Mathur, "Ransomware (Malware) Detection using Machine Learning," Github, [Online]. Available: https://github.com/muditmathur2020/RansomwareDetection/blob/master/RansomwareDetection.ipynb.

11

Detecting High-quality GAN-generated Face Images using Neural Networks

Ehsan Nowroozi[1] and Yassine Mekdad[2]

[1]Faculty of Engineering and Natural Sciences (FENS), Center of Excellence in Data Analytics (VERIÍM), Sabanci University, Turkey
[2]Cyber-Physical Systems Security Lab, Department of Electrical and Computer Engineering, Florida International University, USA
E-mail: ehsan.nowroozi@sabanciuniv.edu; ymekdad@fiu.edu

Abstract

In the past decades, the excessive use of the last-generation Generative Adversarial Network (GAN) models in computer vision has enabled the creation of artificial face images that are visually indistinguishable from genuine ones. These images are particularly used in adversarial settings to create fake social media accounts and other fake online profiles. Such malicious activities can negatively impact the trustworthiness of users' identities. On the other hand, the recent development of GAN models may create high-quality face images without evidence of spatial artifacts. Therefore, reassembling uniform color channel correlations is a challenging research problem. To face these challenges, we need to develop efficient tools able to differentiate between fake and authentic face images.

In this chapter, we propose a new strategy to differentiate GAN-generated images from authentic images by leveraging spectral band discrepancies, focusing on artificial face image synthesis. In particular, we enable the digital preservation of face images using the cross-band co-occurrence matrix and spatial co-occurrence matrix. Then, we implement these techniques and feed them to a convolutional neural network (CNN) architecture to identify the real from artificial faces. Additionally, we show that the performance boost is particularly significant and achieves more than 92% in different post-processing environments. Finally, we provide several research observations demonstrating that this strategy improves a comparable detection method based only on intra-band spatial co-occurrences.

Keywords: Convolutional neural networks, machine and deep learning, generative adversarial networks, DeepFake, adversarial multimedia forensics, adversarial learning, cybersecurity.

11.1 Introduction

The recent advancements of the machine and deep learning models created challenging models referred to as generative adversarial networks (GANs) [7]. These networks enable the creation of totally artificial images from scratch and added extras such as image editing, attribute manipulation, and style transfer. Apparently, non-expert humans may generate realistic fake photos because of the remarkable efficiency of deep neural networks. In particular, the most newer versions of GAN networks can produce very high-quality pictures, such as facial images, which may effectively fool the human user [2]. Given several real-world applications, the potential misuse of GAN synthetic content poses a significant concern, thus necessitating the creation of image forensics systems capable of distinguishing between genuine and artificial images. Different approaches have been suggested in multimedia forensics to identify whether an image is real or GAN-generated. The most modern approaches are focused on CNN models and can achieve outstanding results. In [16], the authors proposed an approach that can produce extremely excellent detection performance of GAN-generated images. The approach generates the co-occurrence matrix from different color image bands, and then feeds them into the convolutional neural network (CNN).

Observing that recent GAN models may create extremely high-quality images with virtually imperceptible spatial errors (if any), reconstructing a consistent relationship between colors is likely more challenging. In this chapter, we enhance the recognition of GAN-produced images by expanding the techniques presented in [16]. More specifically, we supply the CNN detection method using cross-band and gray-level co-occurrences that are estimated individually on the single-color bands. Our experimental results rely on the well-known StyleGAN model [9], [10] that creates higher quality images than prior models such as ProGAN [8], attempting to make the detection task more difficult. In comparison to the intra-band technique suggested in [16], our CNN detector achieves almost optimal detection performance, which has far higher robustness to post-processing.

Given the vulnerability of several approaches regarding the recognition of JPEG compressed faces, we re-trained the model using JPEG compressed faces to recognize JPEG-compressed artificial images. Our experimental results demonstrated that the JPEG-aware models could perfectly detect with high level of accuracy the JPEG compressed GAN images in matching and mismatching scenarios. In other words, the JPEG quality factors are different when they are used only for training and validation. In this study, our experimental results exhibit the effectiveness of the JPEG-aware models in the post-processing environments performed to the images previously to JPEG compression, thus confirming that the JPEG-aware versions of the detectors are particularly resistant versus post-processing.

11.1.1 Organization

The rest of our chapter is organized as follows. We briefly cover research efforts on GAN image recognition in Section 11.2, emphasizing relevant approaches to color space analysis. In Section 11.3, we present the review and approval techniques. Afterward, we describe our approach in Section 11.4. Then, we provide and analyze the outcomes of our results in Section 11.5. Finally, we conclude our chapter with some last considerations in Section 11.6.

11.2 State of the Art

Prior works have suggested several methods to differentiate between artificial images generated through GANs and genuine ones. Some algorithms use particular facial traces [14]. For instance, the authors in [18] demonstrated that by evaluating the placements of facial feature spots and using them for training an SVM, we can identify GAN-generated images. Other approaches that utilize color information to disclose GAN features are more similar to the one given in [15]. In this approach, the authors presented two metrics based on comparisons among color information and intensity for the detection by examining GAN networks' behavior and color cues.

In [1] and [3], the co-occurrence features, which are frequently calculated on residual images, are commonly employed in detection techniques for identifying or localization modifications. These approaches commonly consider the SPAM feature [17], the rich feature model [5], and the rich features modeling techniques for color images [6], which were first developed for image steganography. Additional studies have been suggested by [11], and the authors incorporated the assessment of the color channel and the SPAM-like properties to detect GAN images. In this case, the co-occurrences matrices are retrieved from truncated residuals of numerous color components and a truncated residual image constructed by merging the Red, Green, and Blue bands. Afterward, the co-occurrence matrices are concatenated into a feature set and then used to train the SVM.

Convolutional neural networks (CNN) based approaches have recently been suggested by [12] and [13]. These approaches outperform prior techniques based on fundamental machine learning and handcrafted features. In [13], the authors presented an incremental learning strategy to alleviate the necessity for large trained data from all the various GAN models. In a recent paper [16], the authors demonstrated that providing as input the co-occurrence matrix (that is performed on the source images) to the CNN model can significantly improve the face GAN identification challenge when compared to co-occurrences extracted features from noise residuals.

In this chapter, we benefit from the increased effectiveness of the CNN model and develop a methodology to detect dissimilarities among the color components by observing the co-occurrence cross-band matrices.

11.3 Cross Co-occurrences Feature Computation

It was recently demonstrated that discrepancies in pixel co-occurrences might reveal GAN content. In [16], the authors recommended that the three co-occurrence matrices generated on the Red, Green, and Blue channels of the images can be fed into a CNN as a tensor of three matrices. Afterward, the network will learn unique characteristics from the co-occurrence matrices by learning with GAN and real images, thus outperforming state-of-the-art approaches. Accordingly, we refer to *Co-Net*, the CNN model that is trained on co-occurrence matrices.

In this study, we observe that recreating consistent relationships across color bands is a challenging task for GANs. Therefore, to leverage the correlations between color bands, we consider estimating cross-co-occurrences and feeding them into the CNN model simultaneously with the spatial co-occurrences estimated on the single bands. In this case, we denote *Cross-Co-Net* as the name given to the network that has been trained using this strategy. It is worth mentioning that cross-band features should be resistant to the standard post-processing operations that are stressing on spatial pixel correlations instead of cross-band features.

Given an image $I = (a, b, c)$ of size $H \times V \times 3$, we note Red, Green, and Blue the 2D matrix channels, respectively. For a displacement $\tau = (\tau_a, \tau_b)$, we define the spatial co-occurrence matrix of channel Red, Green, and Blue as follows:

$$W_\tau(x, y; Red) = \sum_{a=1}^{H} \sum_{b=1}^{V} \begin{cases} 1 & if\ \tau(a, b) = x\ and\ Red(a + \tau_a, b + \tau_b) = y \\ 0 & otherwise \end{cases} \tag{11.1}$$

$$W_\tau(x, y; Green) = \sum_{a=1}^{H} \sum_{b=1}^{V} \begin{cases} 1 & if\ \tau(a, b) = x\ and\ Green(a + \tau_a, b + \tau_b) = y \\ 0 & otherwise \end{cases} \tag{11.2}$$

$$W_\tau(x, y; Blue) = \sum_{a=1}^{H} \sum_{b=1}^{V} \begin{cases} 1 & if\ \tau(a, b) =\ and\ Blue(a + \tau_a, b + \tau_b) = y \\ 0 & otherwise \end{cases}. \tag{11.3}$$

For Red, Green, and Blue channels, $I(a, b, 1)$, where x, y are integers between 0 and 255. We note that *RG, GB, and RB* stand for *Red and Green*, *Green and Blue*, and *Red and Blue*, respectively. In what follows, we describe our construction of the cross-co-occurrence matrix (spectral) for the channels Red, Green, and Blue:

$$W_{\tau'}(x, y; RG) = \sum_{a=1}^{H} \sum_{b=1}^{V} \begin{cases} 1 & if\ I(a, b, 1) = x\ and\ I(a + \tau'_a, b + \tau'_b, 2) = y \\ 0 & \text{otherwise} \end{cases} \tag{11.4}$$

$$W_{\tau'}(x, y; GB) = \sum_{a=1}^{H} \sum_{b=1}^{V} \begin{cases} 1 & if\ I(a, b, 1) = x\ and\ I(a + \tau'_a, b + \tau'_b, 2) = y \\ 0 & \text{otherwise} \end{cases} \tag{11.5}$$

$$W_{\tau'}(x, y; RB) = \sum_{a=1}^{H} \sum_{b=1}^{V} \begin{cases} 1 & if\ I(a, b, 1) = x\ and\ I(a + \tau'_a, b + \tau'_b, 2) = y \\ 0 & \text{otherwise} \end{cases}. \tag{11.6}$$

Rather than being given intra-channel, the offsets $\tau' = (\tau'_a, \tau'_b)$ are used inter-channel–in other words, across the bands. The offset in the cross-band scenario accounts for two impacts: the difference band as well as the different spatial position. For cross-co-occurrence analyses, a τ' of [0, 0] represents the condition where there is not any movement in the image pixel; on the other hand, there is a movement in spectral bands that we considered pixel conditions [1, 1]. The six tensors that comprise the $Cross\text{-}Co\text{-}Net$ network's

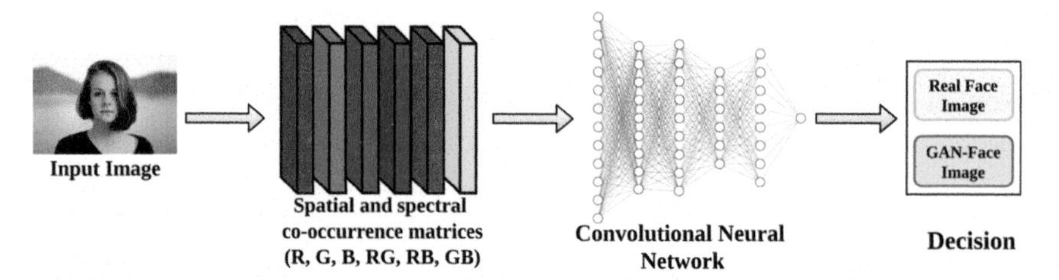

Figure 11.1 The considered CNN-based detector scheme for high-quality GAN-generated face images

($F_{T,T'}$) input are the three co-occurrence matrices for the channels [Red, Green, and Blue] and the three cross-co-occurrence matrices for the couple [Red Green, Red Blue, and Green Blue], namely:

$$F_{T,T'}(x,y;:) = [W_T(x,y;Red), W_T(x,y;Green), W_T(x,y;Blue), \\ W_{T'}(x,y;Red\ and\ Green), W_{T'}(x,y;Red\ and\ Blue), \\ W_{T'}(x,y;Green\ and\ Blue)]. \quad (11.7)$$

In Figure 11.1, we present our proposed CNN detector architecture for high-quality GAN-generated face images. First, we input the source image to the spatial and spectral co-occurrence matrices. Then, we output the corresponding features to the convolutional neural network, which will eventually decide whether the input source image is a real face image or a GAN-face image.

11.4 Evaluation Methodology

In the absence of malicious attackers, we anticipate a good performance from the Cross-Co-Net and Co-Net detectors. Here, we refer to the resiliency of the network. Moreover, we improve the security of the classifiers by retraining them with the attacks that degrade their effectiveness. In this section, we present the experimental procedures to analyze the performance of our approach.

11.4.1 Datasets

We examined large real-world datasets with legitimate and malicious facial images, namely, StyleGan2 [10] and VIPPrint datasets. Interestingly, both datasets are considered high-quality and challenging datasets.

StyleGan2 Dataset: The StyleGAN2 image dataset [10] has recently been suggested as a refinement to the original StyleGAN design [9] and has the capability of producing outstanding results, and thus by generating exceedingly high-quality artificial images. Additionally, face images developed using StyleGAN2 are hardly distinguishable from authentic faces (without visual effects on the face and backgrounds area), making their identification extremely difficult. To train the StyleGAN dataset, we use the Flicker-Faces-HQ (FFHQ) database. It consists of a set of 70,000 high-quality images of human faces created as a standard for GAN networks. These images are provided in PNG image format with a size of 1024×1024 and illustrate a wide range of ages, genders, races, and so forth. The dataset, which includes faces with accessories like eyeglasses, sunglasses, hats, and so on, is automatically cropped and aligned. Figure 11.2 illustrates different natural and generated StyleGan2 facial images.

VIPPrint Dataset: VIPPrint image dataset [4] is primarily focused on evaluating artificial image identification and source linkage algorithms on a one-of-a-kind large dataset of printouts. VIPPrint's research is mostly focused on recognizing a particular printer's identity. Each printer incorporates one-of-a-kind effects into its printed colored materials. The VIPPrint approach recognizes these impacts and employs them in developing a validation system. The primary set of face image examples from the FFHQ dataset was considered to generate VIPPrint images [9]. In this dataset, a total number of 40,000 Kyocera TaskAlfa3551ci face images have been printed and scanned.

(a) Authentic FFHQ images (b) StyleGAN2-generated images

Figure 11.2 Illustrations of natural FFHQ and StyleGAN2-generated images that are hardly distinguishable

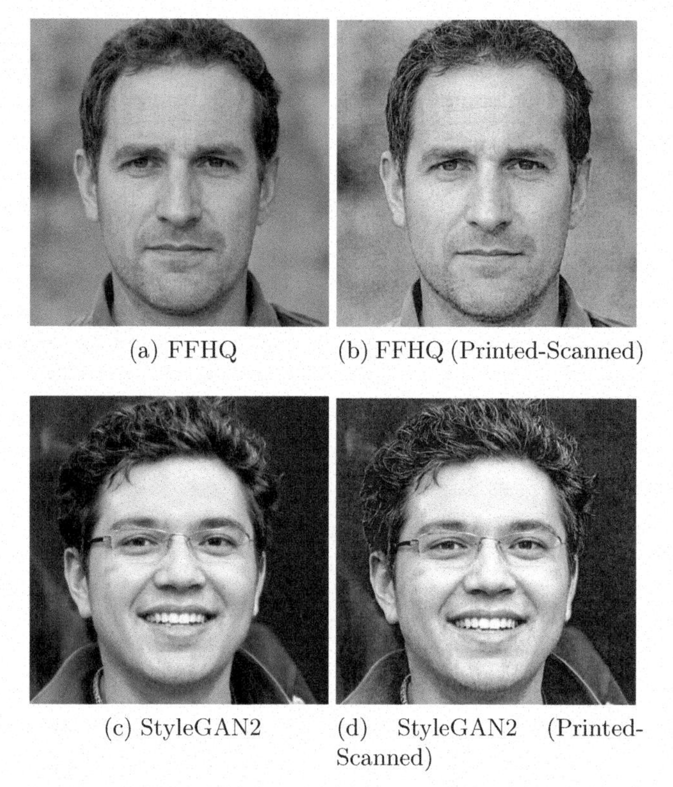

Figure 11.3 Illustrations of authentic FFHQ and StyleGAN2 generated face images with their printed-scanned versions

Figure 11.3 illustrates the source and generated StyleGan2 digital, printed, and scanned versions.

11.4.2 Network Architecture

In the network architecture, we employed the same network structure as in [16], which was previously utilized to recognize GAN images. In particular, we used six convolutional layers followed by a single fully connected layer in the network (we refer to as Cross-Co-Net). The only difference we performed was the first input because we had a six-band input rather than three in our scenario (we refer to as Co-Net). In Figure 11.4, we illustrate the pipeline of our network architecture, and we describe its structure as follows[1]:

[1]For each convolution layer, the stride is fixed to one.

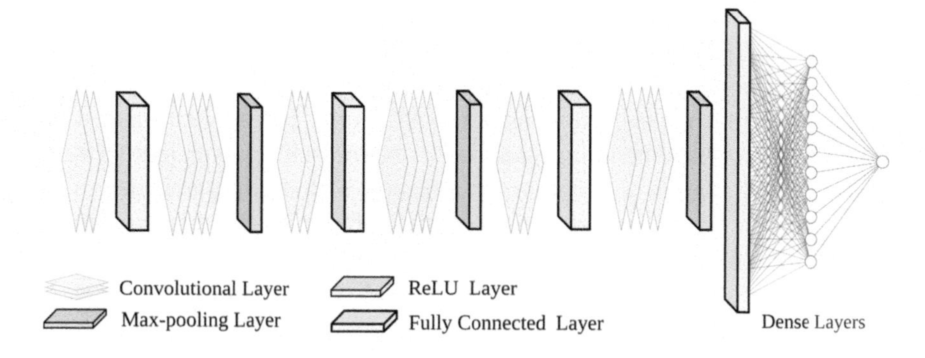

Figure 11.4 Pipeline of the proposed network architecture

- The first convolutional layer has 32 filters of size 3×3, then followed by a ReLu layer.
- The second convolutional layer has 32 filters of size 5×5, then followed by a max pooling layer.
- The third convolutional layer has 64 filters of size 3×3, then followed by a second ReLu layer.
- The fourth convolutional layer has 64 filters of size 5×5, then followed by a second max pooling layer.
- The fifth convolutional layer has 128 filters of size 3×3, then followed by a third ReLu layer.
- The sixth convolutional layer has 128 filters of size 5×5, then followed by a third max pooling layer.
- Finally, a dense layer with 256 nodes and followed by a sigmoid layer.

11.4.3 Resilience Analysis

We evaluated the effectiveness of resiliency against several post-processing techniques. In particular, we compared the Cross-Co-Net with the network proposed in [16]. Later, we deployed geometric manipulations such as resizing, rotating, zooming, cropping, and filtering operations (e.g., median filtering, blurring, and contrast enhancements such as gamma correction and adaptive histogram equalization (AHE)). In terms of resizing, we downscaled the images to [0.9, 0.8, and 0.5] as scaling factors and upscaled (referred to as zooming) the utilizing factors [1.1, 1.2, and 1.9], with bicubic interpolation. For rotational operations, we employed bicubic interpolation to assess [5, 10, and 45] angles. We performed the cropping by considering 880×880. For the filtering, we chose the window sizes for median filtering and blurring at [3×3, and 5×5], respectively. We additionally ran a test versus Gaussian noise in a real-world scenario applying different standard deviations [0.5, 0.8, and 2] and zero means. We considered additional operations to assess the model's robustness, including gamma correction with parameters [0.8, 0.9, and 1.2] and adaptive histogram equalization with a clip parameter of 1.0. Furthermore, we evaluated the robustness of our model against two techniques that follow after each other, including blurring with window size 3×3 and a kernel [[-1, -1, -1], [-1, 9, -1], [-1, -1, -1]] for sharpening operation.

Our analysis demonstrates that the JPEG-aware version verified the effectiveness of the Cross-Co-Net training technique. It is noteworthy since the attackers often use JPEG compression in digital image forensics to eliminate modification traces. Additionally, we evaluated the effectiveness of the JPEG-aware Cross-Co-Net detector by considering the compression post-processing procedures. For the cases of resizing with scaling factor 0.9, median filtering with a window size of 3 3, and Gaussian noise with a standard deviation of 2, we created a JPEG compressed version of the processed images, along with the compression performed with various quality factors (QFs).

11.5 Experimental Results

In this section, we present the experimental results obtained to evaluate the effectiveness of our newly suggested technique using two different challenging datasets, StyleGAN2 and VIPPrint.

11.5.1 Experimental Settings

For the experiments, we used 20,000 authentic FFHQ and 20,000 GAN-generated images from the StyleGAN2 dataset, divided as follows for both authentic and GANs: we considered 12,000 datasets for training, 4000 for validation, and 4000 for testing. Regarding the VIPPrint dataset, we divided 40,000 images for both authentics and GANs as follows: we considered 20,000 print-scanned images for training, followed by 10,000 for validation and 10,000 for testing.

Additionally, we used stochastic gradient descent (SGD) as the optimizer, with a learning rate of 0.01, the momentum of 0.9, the batch size of 40, and training epochs of 40. We implemented our network using TensorFlow's

Keras API for training and testing. We also trained the Co-Net network with the same settings to guarantee a fair evaluation. We used the OpenCV package in Python to post-processing the robustness of our experiments. We considered a total of 2000 per-class images from the test set for each processing operation and parameter. To identify GAN-generated face images, we constructed the JPEG-aware Cross-CoNet model and utilized the following quality factors [75, 80, 85, 90, and 95]. With 15,000 images, we selected 3000 images for training for each quality factor and 5000 images for validation and testing in the real and GAN classes. Then, we retrained the model over 40 epochs, with SGD as an optimizer. We used a similar set of test images to evaluate the robustness of the JPEG-aware model against post-processing.

11.5.2 Performance and Robustness of the Detector

In this subsection, we evaluate the performance of Cross-Co-Net and Co-Net detectors and their robustness in the existence of different post-processing.

Cross-Co-Net and Co-Net Performance: For the StyleGAN2 detection task from a test set, we remark in Table 11.1 that Cross-Co-Net achieved a test accuracy of 99.80%, which is slightly higher than Co-Net of 98.25% in the unaware scenario. Additionally, 99.53% test accuracy was obtained for Cross-Co-Net and 98.60% for Co-Net for the VIPPrint dataset. Overall, the Cross-Co-Net has a significant advantage over Co-Net in post-processing robustness. For the JPEG-aware scenario, the Cross-Co-Net achieves an average accuracy of 94.40% on the styleGAN2 datatset and an accuracy rate of 93.80% by retraining JPEG-aware versions of the Co-Net model, with a small improvement of Cross-Co-Net over Co-Net that was similar. The JPEG-aware Cross-Co-Net for a VIPPrint dataset acquires an average accuracy of 93.05% on the test set for JPEG genuine and fake faces with similar quality characteristics. For the JPEG-aware of Co-Net using a VIPPrint dataset, the training resulted in a test accuracy of 92.56%.

Table 11.1 Accuracies of Cros-Co-Net and Co-Net for unaware and JPEG-aware scenarios over StyleGAN2 and VIPPrint datasets

Scenario	Network	StyleGAN2 dataset	VIPPrint dataset
Unaware scenario	Cross-Co-Net	99.80%	99.53%
	Co-Net	98.25%	98.60%
JPEG-aware scenario	Cross-Co-Net	94.40%	93.05%
	Co-Net	93.80%	92.56%

Furthermore, we illustrate the differences among spectral and spatial bands in a single real image separately in Figure 11.5. In Figure 11.6, we show the discrepancies in a single GAN StyleGAN2 image.

Cross-Co-Net and Co-Net Robustness: We present the accuracies of the tests performed in different post-processing operations in Table 11.2 by considering StyleGAN2 and VIPPrint datasets. We observe that Cross-Co-Net achieves substantially more heightened robustness in all cases, even if the image post-processing is strong. We also remark that the worst-case situation refers to AHE and blurring followed by sharpening, where the Cross-Co-Net accuracy decreases to 75% or slightly less. According to the Co-Net's results, we can notice that the network's accuracy is close or equal to 50% in many scenarios. The rationale for such a result is that nearly all GAN images are labeled as real after post-processing, implying that the artifacts on which the model turns to detect the GAN-image are removed through post-processing. On the other hand, the Cross-Co-Net performs well despite a small loss in performance, demonstrating that by analyzing cross-band co-occurrences, the network may learn in-depth features and characteristics of GAN images, resulting in a robust model subsequent processing. Furthermore, most processing operations modify the spatial relations

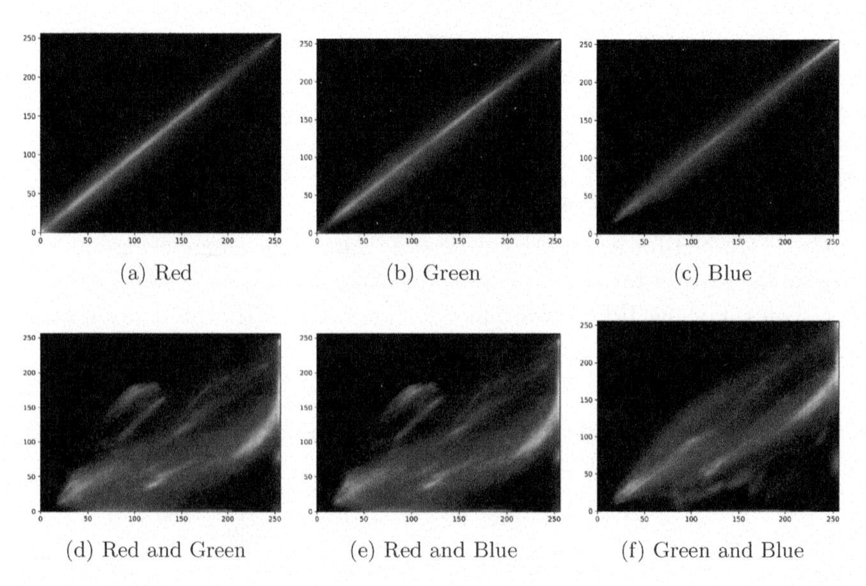

(a) Red (b) Green (c) Blue

(d) Red and Green (e) Red and Blue (f) Green and Blue

Figure 11.5 Visual representation of the Co-occurrence matrices for real image with different channel combinations

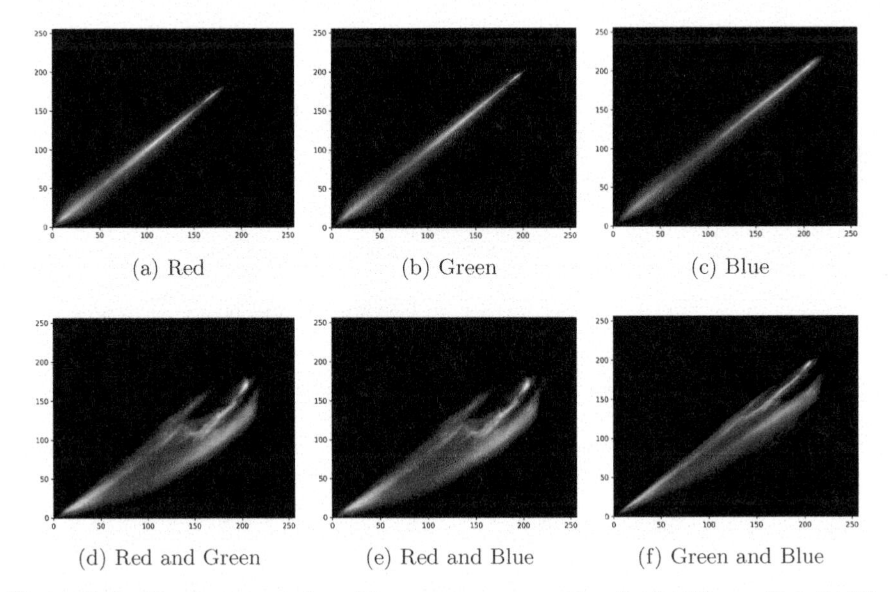

(a) Red (b) Green (c) Blue

(d) Red and Green (e) Red and Blue (f) Green and Blue

Figure 11.6 Visual representation of the co-occurrence matrices for GAN image StyleGAN2 with different channel combinations

across pixels but not the intra-channel relations, providing another reason for Cross-Co-Net's robustness versus post-processing.

11.5.3 Performance and Robustness of JPEG-aware Cross-Co-Net

In this subsection, we present the effectiveness and robustness of JPEG-aware Cross-Co-Net in the match and mismatch JPEG quality factors.

Performance under different quality factors: As an additional study, we verified that the efficiency of both Cross-Co-Net and Co-Net degrades when JPEG compression is used. Particularly, with QF = [95], detection performance is already less than 90%, falling under 80% when QF = [85]. Nevertheless, the loss of efficiency using JPEG compression is not unexpected. The majority of authentic images in the FFHQ dataset have been at least once JPEG compressed, thus producing compression footprints. On the other hand, the GAN images do not show such traces. The network may then implicitly associate compression artifacts with the class of actual images. According to our results, the JPEG-aware Cross-Co-Net achieves an average accuracy of 94.40 percent on a test set of JPEG authentic and GAN

Table 11.2 Robustness performances in the presence of post-processing for StyleGAN2 and VIPPrint dataset

		StyleGAN2 dataset		VIPPrint dataset	
Operations	**Parameters**	**Cross-Co-Net**	**Co-Net**	**Cross-Co-Net**	**Co-Net**
Median filter	3 × 3	96.25%	50.00%	95.15%	50.00%
	5 × 5	90.35%	50.00%	89.25%	50.00%
Gaussian noise	0.5	99.95%	86.40%	97.75%	83.10%
	0.8	99.55%	64.10%	97.87%	61.20%
	2	90.70%	50.00%	87.60%	50.00%
AHE	-	75.00%	50.00%	72.00%	50.00%
Gamma correction	0.9	99.65%	55.90%	97.60%	52.85%
	0.8	82.50%	50.80%	82.30%	50.00%
	1.2	91.70%	50.15%	90.63%	50.05%
Average blurring	3 × 3	92.85%	72.50%	90.80%	67.00%
	5 × 5	85.30%	54.10%	81.00%	50.10%
Resizing	0.9	99.73%	90.78%	97.70%	85.91%
	0.8	99.50%	76.65%	98.50%	73.60%
	0.5	81.50%	50.05%	80.00%	50.00%
Zooming	1.1	99.60%	94.95%	99.00%	90.05%
	1.2	99.45%	89.95%	97.10%	84.00%
	1.9	98.60%	57.65%	96.00%	52.30%
Rotation	5	99.45%	93.65%	99.00%	90.15%
	10	99.50%	93.65%	98.50%	91.60%
	45	99.50%	71.90%	98.50%	70.09%
Cropping	-	99.80%	92.60%	97.60%	91.32%
Blurring followed by sharpening	-	73.60%	50.00%	73.00%	51.10%

images when the quality factors [75, 80, 85, 90, and 95] are used for training. Therefore, the model trained on the stated QFs values can be generalized to other compression settings. Table 11.3 presents the outcomes of tests run on a StyleGAN2 dataset containing both match and mismatch QF values. When mismatch QFs are taken into account, the actual loss of accuracy in the ranges [73, 77, 83, 87, 93, and 97] is less than 1%, suggesting that the network's generalization capabilities are rather impressive. Table 11.3 also shows the average loss of accuracy for match and mismatch QF values in the VIPPrint dataset. While VIPPrint dataset detection is significantly more complicated in this case, the average accuracy loss is less than 3%.

Robustness under different quality factors: We proved that the efficiency generated by the JPEG-aware Cross-Co-Net network is quite small when processing operations are performed before the final compression. For the scenarios of median filtering with a window size of 5×5, resizing with a

Table 11.3 Accuracies of the JPEG-aware Cross-Co-Net for matched and mismatched quality factors for StyleGAN2 and VIPPrint dataset

	Accuracy	
Quality factors	**StyleGAN2 dataset**	**VIPPrint dataset**
73	95.40%	93.35%
75	95.83%	93.80%
77	95.93%	93.90%
80	96.35%	96.00%
83	95.73%	93.60%
85	96.28%	93.08%
87	96.50%	96.55%
90	95.70%	93.78%
93	96.60%	96.00%
95	96.10%	96.41%
97	95.80%	93.75%

scaling factor of 0.8, Gaussian noise with a standard deviation of 2, zooming with a scaling factor of 1.9, and AHE, we report the results in Tables 11.4, and 11.5, by considering StyleGAN2 and VIPPrint, respectively. This study reveals that the JPEG-aware edition of Cross-Co-Net maintains the model's sufficient robustness, indicating that concentrating on color band discrepancies via cross-band co-occurrences is an efficient method to discriminate between authentic and GAN-generated images.

Table 11.4 JPEG-aware Cross-Co-Net robustness performance with the post-processing operators (StyleGAN2).

QFs	Median Filtering	Resizing	Gaussian Noise	Zooming	AHE
73	93.80%	92.10%	93.00%	92.00%	91.00%
75	94.10%	93.10%	92.80%	94.80%	92.70%
77	94.20%	93.70%	93.30%	94.30%	92.20%
80	94.00%	94.40%	93.50%	94.50%	93.60%
83	94.00%	94.00%	93.01%	94.50%	92.05%
85	93.70%	94.10%	93.40%	93.40%	92.00%
87	94.10%	94.40%	93.40%	94.40%	92.00%
90	94.10%	94.40%	93.10%	94.10%	92.10%
93	94.10%	94.30%	91.10%	94.10%	90.10%
95	94.10%	94.80%	88.30%	94.30%	86.04%
97	94.40%	94.60%	88.05%	94.06%	86.00%

Table 11.5 JPEG-aware Cross-Co-Net robustness performance with the post-processing operators (VIPPrint).

QFs	Median Filtering	Resizing	Gaussian Noise	Zooming	AHE
73	92.30%	90.60%	90.09%	90.80%	90.25%
75	93.00%	91.70%	90.50%	92.72%	91.00%
77	93.14%	91.40%	91.10%	92.37%	91.10%
80	93.05%	92.70%	91.70%	92.57%	92.50%
83	93.18%	92.20%	91.00%	92.58%	91.55%
85	92.40%	92.00%	91.30%	91.44%	91.00%
87	93.00%	92.20%	91.30%	92.44%	91.00%
90	93.00%	92.80%	91.00%	92.20%	91.02%
93	93.00%	92.70%	90.00%	92.17%	87.00%
95	93.00%	92.50%	83.20%	92.33%	84.44%
97	93.21%	92.50%	83.15%	92.00%	84.05%

11.6 Conclusion and Future Works

In this chapter, we proposed a CNN method for detecting high-quality GAN-generated images, emphasizing fake face identification. Our suggested approach makes use of spectral band discrepancies and pixel co-occurrence matrices. Furthermore, we used cross-band co-occurrence matrices to construct the CNN architecture for extracting discriminant features for the authentic and GAN classes. The experimental results demonstrate the performance of the proposed approach. Finally, we demonstrated the resilience of our approach to post-processing, compared to single spatial co-occurrences to train the detector, thus illustrating the relevance of considering color band correlations.

In the upcoming work, research effort will be devoted to Cross-Co-Net's performance in both a white-box and a black-box scenario when the band relationships are purposefully changed to mislead the detectors. In this case, it is important to assess the network's effectiveness against a knowledgeable adversary who performs adversarial examples against the CNN model to estimate the co-occurrence computation and backpropagate the gradients to the pixel domain.

References

[1] Barni, M., Nowroozi, E., Tondi, B.: Higher-order, adversary-aware, double jpeg-detection via selected training on attacked samples. In: 2017 25th European Signal Processing Conference (EUSIPCO). pp. 281–285. IEEE (2017)

[2] Brock, A., Donahue, J., Simonyan, K.: Large scale GAN training for high fidelity natural image synthesis. arXiv preprint arXiv:1809.11096 (2018)

[3] Cozzolino, D., Gragnaniello, D., Verdoliva, L.: Image forgery detection through residual-based local descriptors and block-matching. In: 2014 IEEE International Conference on Image Processing (ICIP). pp. 5297–5301. IEEE (2014)

[4] Ferreira, A., Nowroozi, E., Barni, M.: Vipprint: Validating synthetic image detection and source linking methods on a large scale dataset of printed documents. Journal of Imaging **7**(3), 50 (2021)

[5] Fridrich, J., Kodovsky, J.: Rich models for steganalysis of digital images. IEEE Transactions on information Forensics and Security **7**(3), 868–882 (2012)

[6] Goljan, M., Fridrich, J., Cogranne, R.: Rich model for steganalysis of color images. In: 2014 IEEE International Workshop on Information Forensics and Security (WIFS). pp. 185–190. IEEE (2014)

[7] Goodfellow, I., Pouget-Abadie, J., Mirza, M., Xu, B., Warde-Farley, D., Ozair, S., Courville, A., Bengio, Y.: Generative adversarial nets. Advances in neural information processing systems **27** (2014)

[8] Karras, T., Aila, T., Laine, S., Lehtinen, J.: Progressive growing of gans for improved quality, stability, and variation. arXiv preprint arXiv:1710.10196 (2017)

[9] Karras, T., Laine, S., Aila, T.: A style-based generator architecture for generative adversarial networks. In: Proceedings of the IEEE/CVF Conference on Computer Vision and Pattern Recognition. pp. 4401–4410 (2019)

[10] Karras, T., Laine, S., Aittala, M., Hellsten, J., Lehtinen, J., Aila, T.: Analyzing and improving the image quality of stylegan. In: Proceedings of the IEEE/CVF Conference on Computer Vision and Pattern Recognition. pp. 8110–8119 (2020)

[11] Li, H., Li, B., Tan, S., Huang, J.: Identification of deep network generated images using disparities in color components. Signal Processing **174**, 107616 (2020)

[12] Mansourifar, H., Shi, W.: One-shot gan generated fake face detection. arXiv preprint arXiv:2003.12244 (2020)

[13] Marra, F., Saltori, C., Boato, G., Verdoliva, L.: Incremental learning for the detection and classification of gan-generated images. In: 2019 IEEE International Workshop on Information Forensics and Security (WIFS). pp. 1–6. IEEE (2019)

[14] Matern, F., Riess, C., Stamminger, M.: Exploiting visual artifacts to expose deepfakes and face manipulations. In: 2019 IEEE Winter Applications of Computer Vision Workshops (WACVW). pp. 83–92. IEEE (2019)

[15] McCloskey, S., Albright, M.: Detecting gan-generated imagery using color cues. arXiv preprint arXiv:1812.08247 (2018)

[16] Nataraj, L., Mohammed, T.M., Manjunath, B., Chandrasekaran, S., Flenner, A., Bappy, J.H., Roy-Chowdhury, A.K.: Detecting GAN generated fake images using co-occurrence matrices. Electronic Imaging **2019**(5), 532–1 (2019)

[17] Pevny, T., Bas, P., Fridrich, J.: Steganalysis by subtractive pixel adjacency matrix. IEEE Transactions on information Forensics and Security **5**(2), 215–224 (2010)

[18] Yang, X., Li, Y., Qi, H., Lyu, S.: Exposing GAN-synthesized faces using landmark locations. In: Proceedings of the ACM Workshop on Information Hiding and Multimedia Security. pp. 113–118 (2019)

12

Fault Tolerance of Network Routers using Machine Learning Techniques

Harinahalli Lokesh Gururaj[1,*], Francesco Flammini[2], Beekanahalli Harish Swathi[1], Nandini Nagaraj[1], and Sunil Kumar Byalaru Ramesh[1]

[1]Department of CSE, Vidyavardhaka College of Engineering, India
[2]University of Applied Sciences and Arts of Southern Switzerland (CH), Switzerland
E-mail: gururaj1711@vvce.ac.in; francesco.flammini@supsi.ch; swathibh@vvce.ac.in; nandiningaraj963@gmail.com; sunilkumar.br@vvce.ac.in
*Correspondence

Abstract

Companies are adopting digital technologies at an exponential rate to increase their productivity. To remain competitive, they need to integrate the latest digital innovations into their processes and IT systems with the latest digital innovations such as mobility, data analytics, the cloud, and the Internet of Things (IoT). Network deployment is a matter of seconds through automation. This has created new configuration or design issues. To solve these routing problems, fault tolerance is used, to provide high timeliness and accuracy in a system. This chapter studies fault tolerance problems in a distributed network system, including large-scale networks such as wide area networks (WANs) and metropolitan area networks (MANs). We propose an approach based on machine learning algorithms such as K-nearest neighbor (KNN) and support vector machine (SVM) to solve the fault tolerance problem in scheduled jobs. We used the KNN algorithm to identify the generation of a reset of failed cluster jobs by calculating the energy radius and distance of the jobs. The SVM is used to classify the fault tolerance jobs by considering the hybrid, core, and single node to analyze the failure of the job at different levels with the respective nodes. The results prove the effectiveness of the proposed approach for fault tolerance analysis of network routers.

Keywords: Failure node detector, fault tolerance, network systems, quality of service, distributed network.

12.1 Introduction

Fault tolerance is used in communication systems. The fastest development of the internet technologies, the mechanism of fault tolerance, is required to provide high opportunity and high accuracy in a system. And it is particularly prominent in distributed network systems. In general, it is important in a large-scale environment. Fault tolerance approaches can be classified into two types: proactive and reactive. Proactive approaches predict errors, faults, and failures and replace the suspected components. Reactive approaches reduce the effect of faults by taking necessary actions. Some fault treatment policies can also prevent faults from being reactivated. Normally, in a distributed system, users want the system to remain operational despite technical failures, even if some of the participants of these network systems have crashed. With a large number of participants and long running time, the probability that the hosts crash during the execution is inevitable, regardless of the physical reliability of each host. Thus, an effective system must be designed and executed so that the system can seamlessly tolerate a reasonable number of host failures, and the occurrence of a good number of host failures is acceptable. Failure detection and process monitoring are the basic components of most techniques for fault tolerance (tolerating failures) in distributed network systems, such as intermediate system to intermediate system (IS-IS), ensemble, transits, and air traffic control systems. Now how to design failure detectors over local networks is a relatively well-known issue, but it is still far from being a solved problem with large-scale systems. Because a large-scale distributed system has lots of difficulties, which need to be addressed if we simulate them as a wired network environment, such as the potentially very large number of monitored processes, the higher probability of message loss, the ever-changing topology of the system, and the high unpredictability of message delays, all the above prominent factors fail to be addressed by the traditional solutions. For effective communication in large-scale distributed

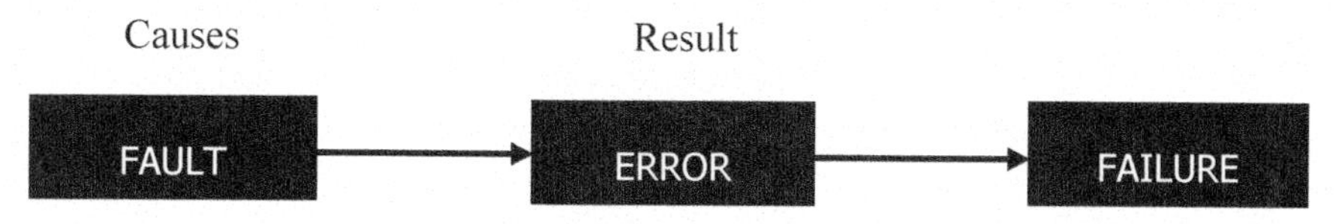

Figure 12.1 Relation diagram of fault error and failure

network systems and its importance, failure detectors should be executed as a typical generic service shared among distributed applications rather than as redundant *ad hoc* network implementations.

If such a generic service can be achieved, it is straightforward to apply failure detectors in any kind of application to ensure the requirement of fault tolerance. Although many ground-breaking advances have been made in failure detection, such as service remains on a distant horizon. There are two main reasons why fault occurs: node failure and malicious error. Different types of failure, such as hardware failure software failure, can be found. Fault tolerance is a setup or configuration that prevents a PC or community device from failing in the event of an unexpected hassle or error, to make a laptop or community fault-tolerant calls for that the client or industrial employer business organization to anticipate how a computer or network tool can also fail and take steps that assist save you that kind of failure.

Communication infrastructure called network-on-chip, for routing the data across the chip it needs a link component and modular router, reactive fault tolerance techniques are commonly used, these systems are affected when the error was occurring [1]. Anomaly detection techniques are compared based on their effectiveness in observing the fault tolerance mechanism, which is based on the anomalies. These anomalies are accurately detected before failure occurs while detecting types of anomalies [25].

Fault tolerance has improved the accuracy and availability of the system. Three necessary terms are used in fault tolerance, i.e., fault, error, and failure, as shown in Figure 12.1 [10]. Figure 12.1 shows that a system is said to fail when it does not fulfill the requirements. An error is part of the system state that may lead to failure or causes of a fault.

12.2 Related Work

In this section, we briefly discuss existing methodologies and their contributions. Wang *et al.* [1] proposed basic transmission for multi-core frameworks on the network which are referred to as network-on-chip (NOC), based on a system-on-chip (SOC) architecture. This technology is exposed in various fields such as fault mechanisms and timing errors. This chapter proposed a proactive fault tolerance technique with reinforcement learning to improve noise performance and fault tolerance. Here, they determine the proactive error approach system for effectively finding an error and revising those errors in a router and processing of transportation. Here, the decision tree algorithm is used. This results in improving noise performance while using different fault strategies. One is selected as optimal among various fault tolerance strategies, reducing the overall network consumption and latency.

Srinivasan *et al.* [2] proposed three phases of the machine learning procedure. In particular, machine-learning-based link fault identification and localization (ML-LFIL) recognizes and restricts connection failure by breaking down the estimations caught from the typical traffic streams, including the total flow rate and packet loss. ML-LFIL works based on three strategies. First, it identifies the disconnection in a node. The second strategy determines the failure of a connection. The third strategy helps to reconnect the node. Here, three algorithms are utilized. The support vector machine helps isolate the data in two different classes by identifying the ideal isolating hyperplane. Multilayer perceptron (MLP), the backpropagation algorithm, is used for training the neural network. Random forest (RF) is a classifier algorithm to construct the decision tree. These algorithms train the learning model based on the mininet platform. The outcome shows that ML-LFIL accomplishes superiority in distinguishing proof and confinement of connection issues with up to 97% precision.

Srinivasan *et al.* [3] proposed traffic engineering based on a machine learning approach to find the current failure in the link node. Machine learning algorithms adopt passive techniques, and a propagation mechanism is used to evaluate the network traffic behavior. The various machine learning algorithms that train the model are as follows. Naive Bayes: it has low complexity during forecasting and training. Logistic regression: here, hypothesis performs between the input features and output variable in the prediction process. Support vector machine: it separates the information and focuses on two classes. Decision tree (DT): it can be classified as non-straight and focuses on different direct surface limits. Random forest: it helps recognize the significant highlights from the informational preparation index. These algorithms are helping to train the learning model in the mininet platform with the iperf3 tool. The results show that a random decision tree algorithm outperforms other algorithms for both topologies with a minimum accuracy in localizing link failures among the examined machine learning algorithms.

Feng *et al.* [4] proposed two models, the NOC architecture model, and the fault model. Network-on-chip architecture is based on the Nostrum [1]. The borderline that separates the network-on-chip architecture from 2D mesh topology is that it connects the output to the input of the same switch. Model faults can be classified into constant fault or defective links in fault. The input port number is similar to the number of output ports in the deflection switch. Connecting all four ties on it, the switch fault is patterned. As long as the network is not dismantled, the terror area can be in any shape. After modeling, the next step is to apply a reconfigurable fault-tolerant deflection routing algorithm. It contains Q routing, fault-tolerant deflection routing algorithm (FTDR), and hierarchical Q-learning deflection routing algorithm (FTDR-H). Q-learning strategy is utilized to reform the routing table by the deflection routing algorithm proposed in the paper to evade errors. It is impervious to the form of a lapse zone. FTDR algorithm and FTDR-H algorithm overshadow other defecting routing algorithms.

Truong-Huu *et al.* [5] proposed a machine learning approach based on traffic engineering (TE) for quick and versatile loss recuperation. The backup path is calculated for each primary path, and the best backup path is selected with that maximum goodness value. It is determined based on three attributes: packet loss ratio, round trip delay, and aggregate flow size apprehended at both source and destination. In this chapter, the proposed model is trained by various machine learning methods such as gradient boosting, linear regression, neural networks, decision tree, random forest, and support vector machine excels. The outcome of the approach proves that loss recuperation time is decreased.

Vindhya *et al.* [6] proposed adaptive routing algorithms with machine learning algorithms for predicting the failure in an NOC. The deficiency in NOC infers inaccessible connections in the system. In those two cases, the connections that can go disconnected are permanent faults, referred to as topological faults and temporary faults: data casualty during the transmission. Here, they deliberate on fault tolerance based on the adaptive routing algorithm. Overcoming this, machine learning algorithms are used to predict the failure in an NOC. The decision tree of a machine learning algorithm is used. For construction, a decision tree is based on the attribute they are temperature and link utilization with the relevant dataset using the ID3 algorithm. The results show that NOC improves its exhibition and network inertness by 30%, based on the typical benchmark.

Wang *et al.* [7] proposed a support vector machine; a double exponential smoothing algorithm is used for detecting errors or failure in an optical network. The main focus is to identify the failure in the optical network and to introduce a decision function from the support vectors in training data monitored by a support vector machine. It is a binary classification algorithm. The actual usage of the algorithm is to find a hyperplane line. In this method, first, the controller selects indicators from collecting the information. The indicators are input optical power, laser bias current, laser temperature offset, usual time, output optical power, and environment temperature. The second is to train the model and the third is to predict the model. The combination of SVM and DES gives high precision of 85%.

Irfan *et al.* [8] proposed fault tolerance redundancy. In the network, identification of node or connection failure infers fault detection. This chapter mainly focuses on communication links. The fault is the origin of interconnection malfunction. The ML paradigm uses the existing information to deliver added importance to decision-making. It effectively utilizes bandwidth and also increases redundancy characteristics of fault tolerance. Also, the complexity of setting up and planning opposes a deterministic and logical answer for this issue and real-time implementation. From standard defined supervised objects, there is an available dataset created by parameter. Packet sniffers such as wire sharks are tools that extract network parameters to identify the most likely errors that occur.

12.2.1 Comparative Analysis of Existing Methodologies

In this section, Table 12.1 shows the performance of existing methodologies. We can see the system's accuracy from the results by considering the literature survey. There are some drawbacks and limitations. Based on the above literature survey, we mention the algorithms and their performance.

12.3 System Architecture

Our proposed system helps identify the fault-tolerant node that fails due to particular faults and is quickly recovered and processed. Here we use machine learning techniques like SVM and KNN algorithms to detect and help recover the nodes. Here, SVM classifies the fault tolerance job or not. Besides, KNN is used to identify the distance of job failure. We energized the fault nodes and helped in further failure avoidance and helped it get going to further process.

Figure 12.2 shows the design of the fault tolerance mechanism. Whenever the reference signal like a job is identified, the reference signal is moved to the references controller. Then the controller is diagnosed with the help of

Table 12.1 Comparative analysis of other methodologies

Number	Algorithms	Limitations	Accuracy
[1]	Reinforcement learning	Less accurate result due to less number of datasets taken	75.24%
[2]	ML-LFIL Backpropagation Random forest	They used both the training and testing datasets; hence, accuracy is low	RF – 73%
[3]	Support vector machine Random forest Decision tree	The method gives less accuracy	Overall accuracy ranges from 43% to 82%
[4]	Reinforcement learning FTDR, FTDR-H	Less accurate because the method is based only on training datasets	72%
[5]	Decision tree Linear regression, Neural networks, Decision tree, Random forest, Support vector machine	More time consumption with less accuracy	70%
[6]	Decision tree	Less number dataset is used	Overall accuracy ranges from 60% to 85%

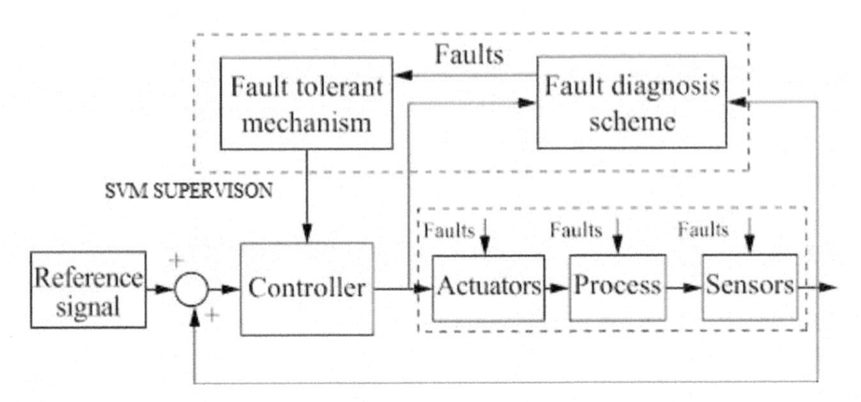

Figure 12.2 Architecture diagram

actuators (processers). Once fault diagnosis is verified, if there are any faults, it will be notified to the fault tolerance mechanism again reinitiated. So, this is a sequential process where the actuators will continuously monitor the job setup processing. And once it fails and reschedules the process, if it is not rescheduled, it gives the output.

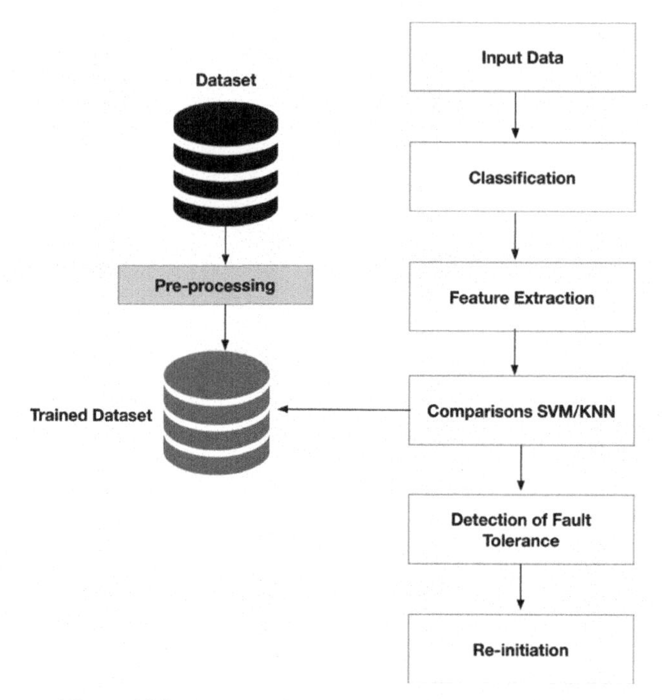

Figure 12.3 Steps performed in fault tolerance detection

Supervised learning is somewhat common in classification problems because it is often the target that makes the computer learn the classification system we created. Recognizing numbers, again, is a common example of learning to classify [21]. In this chapter, a support vector machine and K-nearest neighbor are used.

As shown in Figure 12.3, first, we have to upload the dataset once the datasets are uploaded, and then it has to pre-process, for the generation of training datasets. Next, we consider those datasets as trained datasets. Then segregation is done for testing data. After this, classification is done based on the input data. By extracting features like the amount of support provided by the application based on the SVM and KNN comparison, we discover fault tolerance and then restart the tasks to address them.

12.3.1 Support Vector Machine (SVM)

SVM is a supervised learning algorithm that builds a classifier that maximizes the margin of separation between samples in a training set. It can handle very

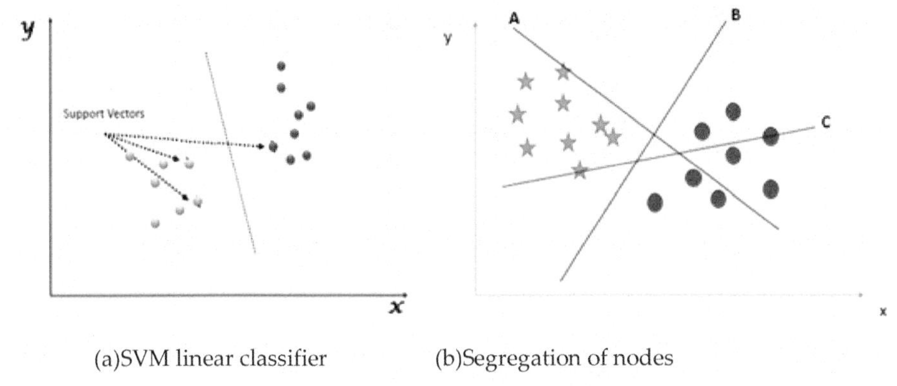

(a)SVM linear classifier (b)Segregation of nodes

Figure 12.4 Example of SVM

large areas of features because SVM training is implemented in this way. The dimensions of the ranked vectors have no distinct effect on the performance of SVM as on the performance of the traditional classifier [16], [17].

This is why it is noted to be particularly effective at high volume classification problems. This will also benefit breakdowns in classification because the number of features it should have is the basis for fault diagnosis that may not be limited. An example of a linear SVM classifier is shown in Figure 12.4. We specifically focus on the binary classification problem in SVM, the most common case, where a dataset has precisely two classes. The primary objective of SVM is to generate a hyperplane function *f*(*x*) between data samples in classes *y*0 (positive samples) and *y*1 (negative samples). Consider a training set.

Figure 12.4(a) depicts an example of a linear classifier (solid line) generated using SVM. The samples that define the boundary (support vectors) are encircled. With samples (*x*1,*y*1),(*x*2,*y*2),(*x*3,*y*3), $\ldots$, (*xn*,*yn*) where $wTx + b \geq 0$ for *di* $=$ +1 and $wTx + b < 0$ for *di* $= --$ 1. The objective of SVM is to find an optimal hyperplane.

The distance of any line is given by $ax + by + c = 0$ from a given point, say, $(x0,\ y0)$ is taken as *d*, the distance of a hyperplane is given as

$$wT\Phi(x) + b = 0.$$

This can be written as

$$d_H\left(\emptyset\left(x_0\right)\right) = \frac{\left|w^t\left(\emptyset\left(x_0\right)\right) + b\right|}{||w||_2} \tag{12.1}$$

where w is a normal vector perpendicular to the hyperplane $f(x)$, and b is a constant. Both w and b are the weighting factors with a maximum distance with the closest sample unless indicated otherwise. We use (ů) as the multiplication operator.

Here, w is the Euclidean norm for the length of w given by

$$||w||_2 =: \sqrt{w_1^2 + w_1^2 + w_1^2 + \cdots + w_n^2}. \tag{12.2}$$

Using Lagrangian multipliers, this problem can be formulated as quadratic programming (QP) optimization problem:

$$L(x, a) = \ f(x) \ \ - - \ ag\ (x),$$

where $\nabla\ (x, a) = 0$.

Partial derivatives w.r.t. x recover the normal parallel constraint.

Partial derivatives λ recovers the $g(x,y) = 0$.

Generally,

$$L(x, a) = f(x) + \sum\_i\ a\_(i\ )\ g\_i\ (x).$$

Lagrangian formula defined as

$$\min L_p \ = |\,|w|\,|^2 - \sum_{i=1}^{l} a_i y_i (x_i\ .\ w + b) \sum_{i=1}^{i} a_i \tag{12.3}$$

Here, α represents Lagrange multipliers for each sample, and L_p indicates the primordial form of the optimization problem.

Algorithm 1: Support Vector Machine [SVM]

Step 1: Collect the fault tolerance dataset features for training
Step 2: Use all the training samples to train and initialize them to SVM
Step 3: Find the best fault tolerance feature for each feature and categorize them
Step 4: Calculate the generalized on each related feature
Step 5: Order the features and find the top percentage of useful features
Step 6: Again, retrain the remaining samples in SVM
Step 7: Binary features classification

12.3.2 K-Nearest Neighbor (KNN)

K-nearest neighbor is a simple but effective method for classification. The main drawbacks regarding KNN are as follows: the first one is low efficiency

- being a lazy learning method, it bans in many applications such as dynamic web mining for a large repository; the second one is its dependence on choosing the best value of *K* [15].

Algorithm 2: *K*-Nearest Neighbor [KNN]

Step 1: Raw data: In this stage, the historical fault tolerance data is agitated from Kaggle or Gitup and this historical data is utilized to predict failure nodes.

Step 2: The next step is data pre-processing. The data pre-processing steps are listed below:

a) Data transformation: Normalization
b) Data cleaning: Fill in missing null values
c) Data integration: Integration of data files

After the dataset is transformed into a pure dataset, the dataset is split into training and testing sets to evaluate. Here, the training values are taken as the more recent values. Testing data is kept as 20% of the total dataset.

Step 3: The user now inputs the relevant data where it is to find and resolve fault tolerance. If it occurs, it is known as a test dataset.

Step 4: Now, we classify the uploaded dataset according to the system needs using the KNN algorithm and after it is ready for the next step, feature extraction. Initialize the *K* value that we take from test dataset, and from the obtained *K* value, we calculate the distance and consider nearest neighbors and make the classification (class).

Step 5: Feature extraction: Only the features to be fed are extracted in this layer. We will choose the feature from the above classification process.

Step 6: Here, the data will be compared with the trained data using KNN and predict the fault tolerance and help the fault nodes to get reinitiated. The data obtained from the classification is mixed, i.e., it is not supervised. Then using KNN, we compare and make clusters to predict and analyze the node failure and help it re-energize.

As shown in Figure 12.5, we have to initialize the *K* value as add value, the range between 20%. Then we calculate the distance of every job that appears in the network. Here, we classify the jobs based on the distance through the radius. Located actuators are identified, and then we reinitiated the job classification for the server.

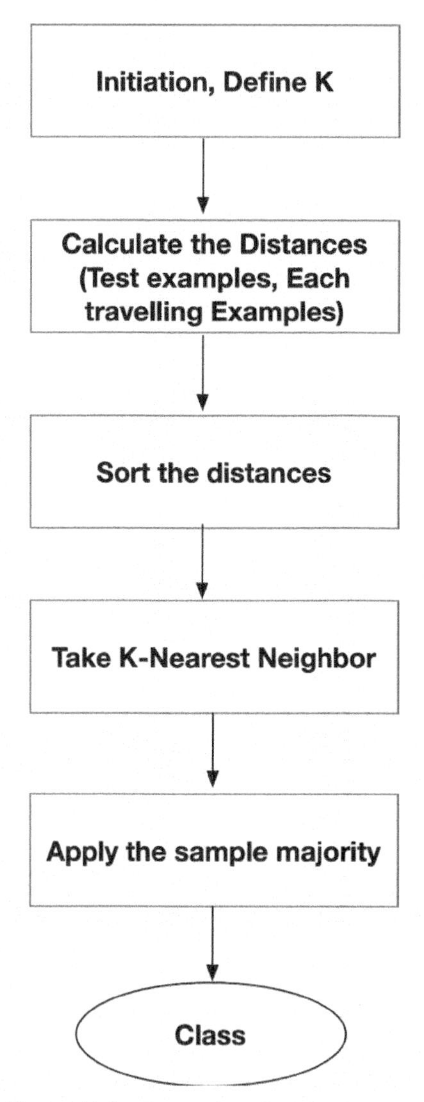

Figure 12.5 Processing of KNN algorithm

We come across the different machine learning algorithms in the existing systems with their performance. Among these, the KNN algorithm gives more accuracy [9]. In our project, we consider both KNN and SVM algorithms to analyze the fault tolerance in the network router. And these two algorithms give an accuracy of up to 95%.

Table 12.2 Experimental setup

Item	Entity used
Platform	Google Colab, Jupyter notebook
Data generated	Yes, data has been generated at campus networks
Experiment	Iterative tests are conducted based on test and training dataset
Pre-processing	Support vector machine (SVM) algorithm was used for data pre-processing

```
*** SVM model training finished ***

iter                    = 8
time cost               = 0.0440 s
obj                     = -0.3568
pData                   = 77.7778 %
nData                   = 22.2222 %
nSVs                    = 3
radio of nSVs           = 6.6667 %
accuracy                = 91.1111 %


*** SVM model test finished ***

time cost               = 0.0040 s
accuracy                = 99.0476 %
```

Figure 12.6 Time cost with an accuracy of hybrid kernel

12.4 Result Analysis

In this section, each figure below represents the result of the proposed methodology. To analyze the fault tolerance of network routers, *K*-nearest neighbor and support vector machine are employed. These algorithms effectively predict the failure in a node.

Figure 12.6 shows the specific iteration level, time cost, estimation for identifying the job process, and the radius frequency of hybrid data. These features are required for the processing of each job as average value and the accuracy for reinitiating the jobs. Finally, it defines a time cost investment for the jobs and the accuracy of a re-initialization of jobs.

Figure 12.7 shows a graphical representation of the radius and distance of every job, which appears in the network and is also based on the job. So, this will help to classify the job, whether it has to reinitiate or not.

Figure 12.8 shows the specific iteration level, time cost, estimation, and accuracy for identifying the job. These facial characteristics are necessary for the subject to a series of actions to achieve a result of each reinitiating the task. After some time, it delimits an opportunity cost investment for the task and the precision or correctness of a re-initialization of a task.

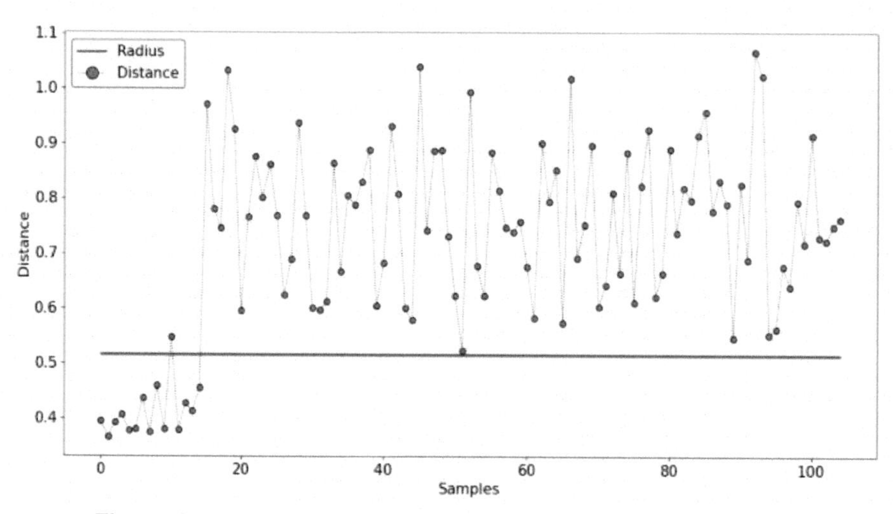

Figure 12.7 Radii with distance calculation using SVM for fault tolerance

```
*** SVM model training finished ***

iter                = 13
time cost           = 0.0312 s
obj                 = -0.7767
pData               = 87.5000 %
nData               = 12.5000 %
nSVs                = 13
radio of nSVs       = 6.5000 %
accuracy            = 97.5000 %


*** SVM model test finished ***

time cost           = 0.0156 s
accuracy            = 96.6667 %

Calculating the grid (0100*0100) scores...

Grid scores completed. Time cost 3.0622 s
```

Figure 12.8 Time cost and accuracy linear kernel

Figure 12.9 indicates the heat map representation because of the jobs that have been initiated in the scheduling of jobs. Here, slots or nodes are defined for the test dataset, training dataset, and support vector configured in the network.

Figure 12.10 specifies the iteration, the time cost, and the accuracy of each node of a single kernel. These elements are required for converting each job as an average value and efficiency for reinitiating the jobs. Finally, it defines a time cost investment for the jobs and the accuracy of a re-initialization of jobs.

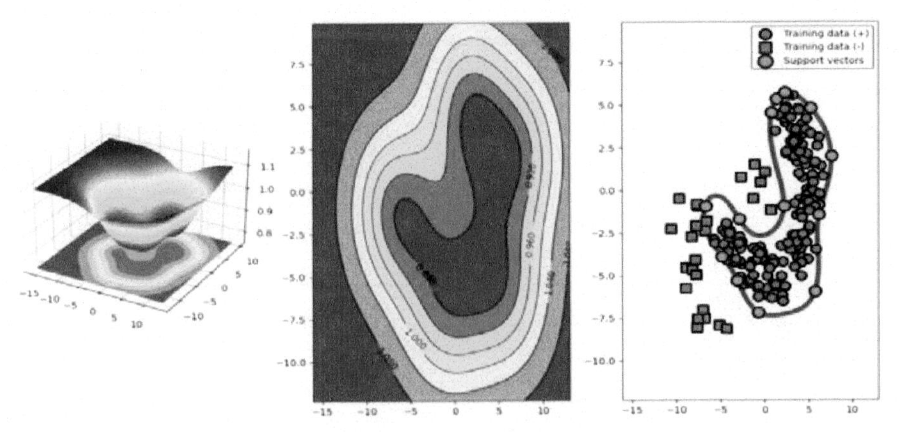

Figure 12.9 Fault tolerance of heat map generation using a kernel

```
*** SVM model training finished

iter                    = 9
time cost               = 0.0469 s
obj                     = -0.7960
pData                   = 100.0000 %
nData                   = 0.0000 %
nSVs                    = 34
radio of nSVs           = 34.0000 %
accuracy                = 99.0000 %


*** SVM model test finished ***

time cost               = 0.0000 s
accuracy                = 96.3542 %
```

Figure 12.10 Time cost and accuracy of the single kernel

Figure 12.11 shows the graphical representation of the radius and distance of every task presented in the network and establishing the job in a single throughput. So, this will categorize the task, whether it has to reinitiate or not.

Figure 12.12 shows the area under the curve (AUC) calculation. It gives the calculation through analysis with the false positive rate (FPR) and the true positive rate (TPR, also called sensitivity). Here, it measures the classifier that can be used to distinguish between the classes that have been generated and the classes that have been terminated during the rescheduling of the jobs.

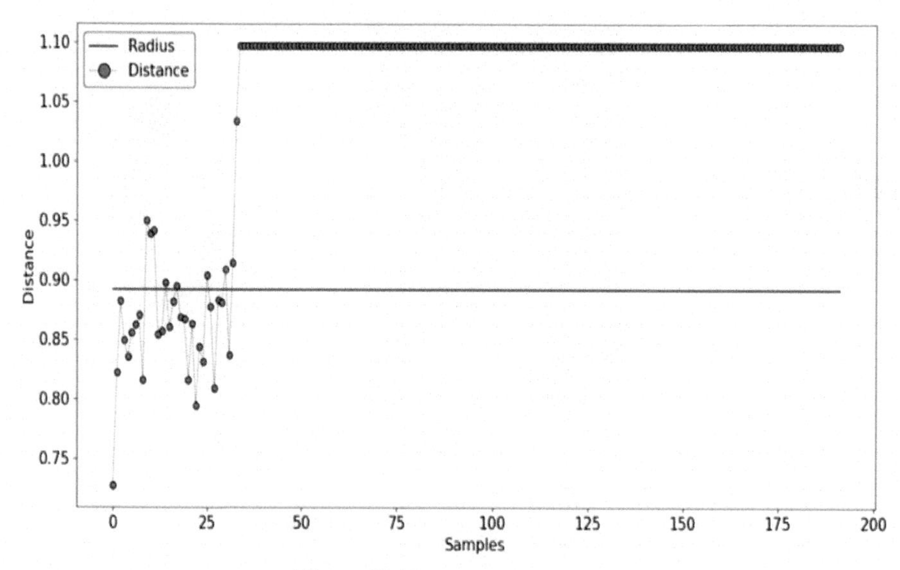

Figure 12.11 Single throughput

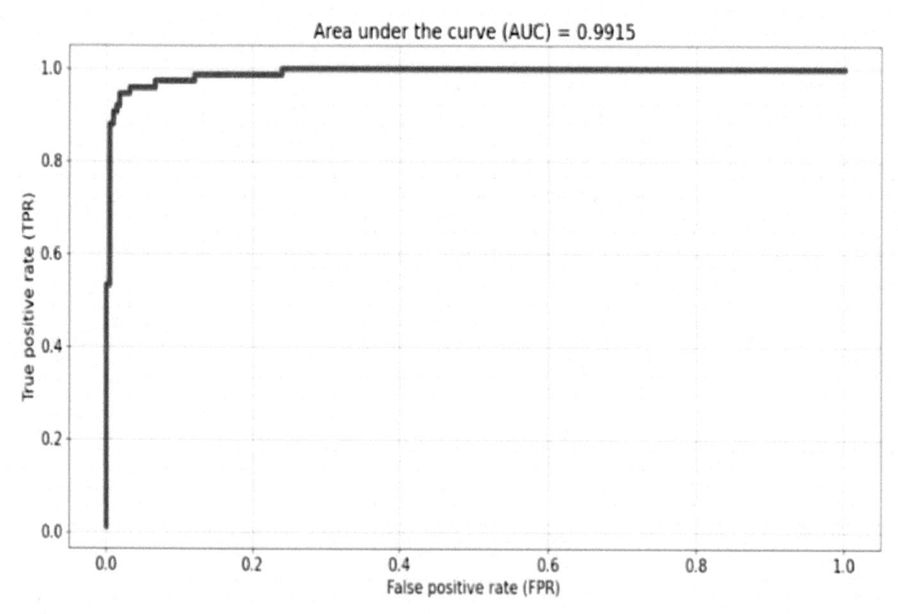

Figure 12.12 AUC calculation of nodes

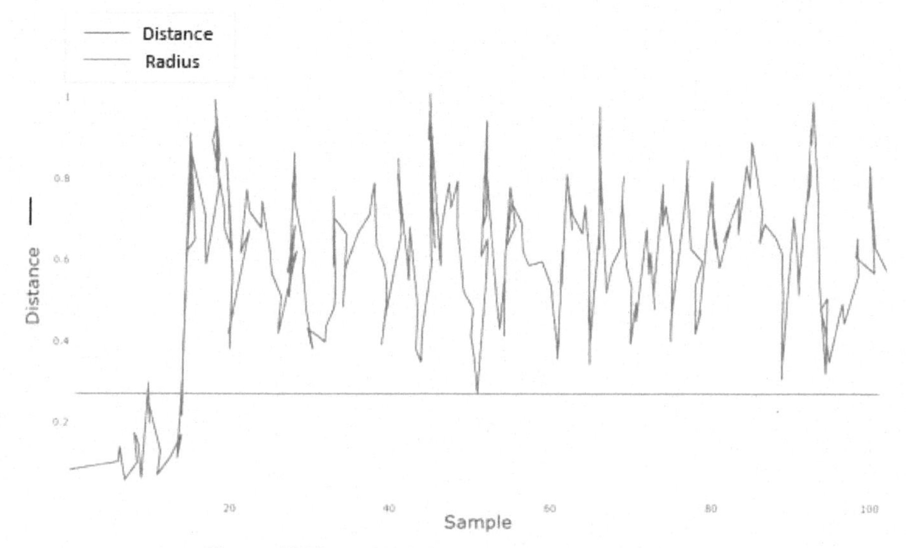

Figure 12.13 Node failure identification using KNN

Figure 12.13 depicts the graphical illustration of the radius and distance of each mission, which is presented within the network primarily based on the KNN model. The operation of KNN is based totally on the category of the nearest one or a few samples. The output of the KNN is the elegance label to the new sample, which is anticipated primarily based on one or extra nearest samples. The KNN effects may additionally deviate as it simply underlines the closest sample.

Figure 12.14 gives a graphical illustration of the radius and distance of each mission, which is presented within the network primarily based on the SVM model. The SVM models can solve problems with non-linear decision boundaries. The kernel function operation is required to identify the faulty nodes. Polynomial, Gaussian, and radian bias are the kernel functions commonly used in SVM.

Figure 12.15 shows the accuracy comparison between SVM and KNN in identifying faulty nodes. SVM and KNN exemplify numerous important exchange-offs in device learning. SVM is less computationally worrying than KNN and is less difficult to interpret. However, it can pick out the most straightforward limited set of patterns. On the other hand, KNN can discover very complex patterns, but its output is tough to interpret. The comparison graph concludes that SVM has achieved swell accuracy compared to KNN.

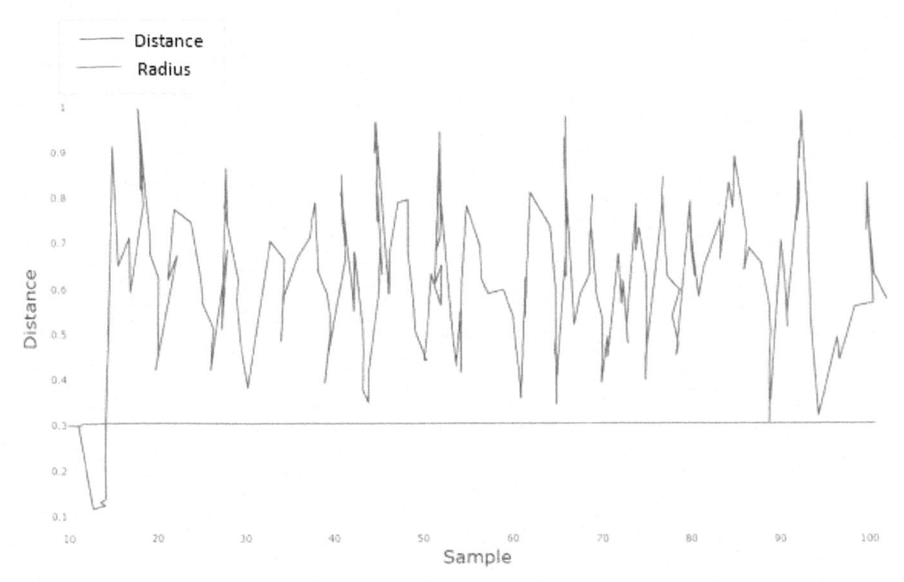

Figure 12.14 Node failure identification using SVM

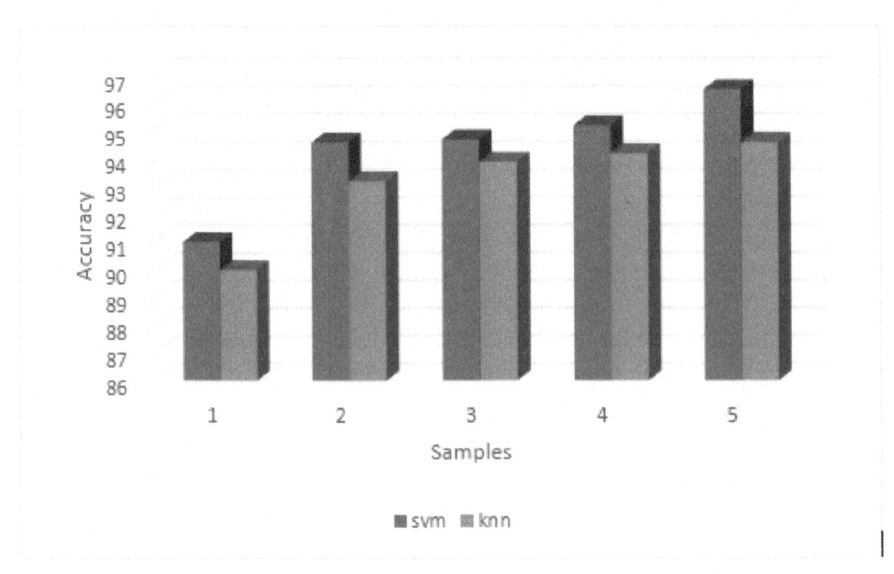

Figure 12.15 Comparison between SVM and KNN

12.5 Conclusion

The proposed methodology uses SVM and KNN approaches to solve the fault tolerance problem in jobs scheduled by considering the energy radius and distance. This chapter suggested methods that solve the problem of the initiation of jobs that are failing at the clusters. We have identified that each layer has its faults and fault tolerance issues. We first explore the relative node fault failure detection, which is an important issue for supporting dependability in distributed systems and often is an important performance bottleneck in the event of node failure. Here, we analyze the failure detection by considering the hybrid, kernel, and single node with an area under the curve (AUC) calculation, time cost using the machine learning technique, and process the failed nodes in a better solution in terms of reliability and energy effectively. This means that to achieve fault tolerance, different aspects and features must be targeted, and no single-focused technology will be able to provide the reliability expected in commercial networks. SVM and KNN are effectively used for fault tolerance analysis of network routers using machine learning techniques.

References

[1] Ke Wang, Ahmed Louri, Avinash Karanth, Razvan Bunescu, "High-Performance, Energy-Efficient, Fault-Tolerant Network-on-Chip Design Using Reinforcement Learning," 2019 Design, Automation & Test in Europe Conference & Exhibition, March 25-29, 2019, Florence, Italy.

[2] Srinikethan Madapuzi Srinivasan, Tram Truong-Huu, "Machine Learning-based Link Fault Identification and Localization in Complex Networks," *IEEE Internet of Things Journal*, vol. 6, no. 4, pp. 6556-6566, 2019.

[3] Srinikethan Madapuzi Srinivasan, Tram Truong-Huu, Mohan Gurusamy, "TE-Based Machine Learning Techniques for Link Fault localization in Complex Networks," *IEEE 6th International Conference on Future Internet of Things and Cloud*, August 6-8, 2018, Barcelona, Spain.

[4] Chaochao Feng, Zhonghai Lu, Axel Jantsch, Jinwen Li, Minxuan Zhang, "A Reconfigurable Fault-Tolerant Deflection Routing Algorithm based on Reinforcement Learning for Network-on-Chip (NOC)," NoCArc'10, December 4, 2010, Atlanta, GA, USA.

[5] Tram Truong-Huu, Prarthana Prathap, Purnima Murali Mohan, Mohan Gurusamy, "Fast and Adaptive Failure Recovery using Machine Learning in Software Defined Networks," *IEEE International Conference on Communications Workshops*, July 11, 2019.
[6] Vindhya N. S., Vidyavathi B. M., "Network on Chip: A Review of Fault Tolerant Adaptive Routing Algorithm," 2018 3rd International Conference on Electrical, Electronics, Communication, Computer Technologies and Optimization Techniques, December 14-15, 2018.
[7] Zhilong Wang, Min Zhang, Danshi Wang, Chung Song, Min Liu, Jin Li, Lioi Lou, Zhuo Liu, "Failure Prediction using Machine Learning and Time Series in Optical Network," *Optics Express*, vol. 25, no. 16, 2017.
[8] Bashir Mohammed, Irfan Awan, Hassan Ugail, Muhammad Younas, "Failure Prediction using Machine Learning in a Virtualized HPC System and Application," *Cluster Computing*, vol. 22, pp. 471-485, 2019.
[9] Guoliang Zhu Kai Lu, Xu Li, Kai Lu, "A Fault-Tolerant K-Means Algorithm based on Storage-Class Memory," *IEEE 4th International Conference on Software Engineering and Service Science (ICSESS)*, May 25, 2013.
[10] Kasem Khalil, Omar Eldash, Ashok Kumar, Magdy Bayoumi, "Machine Learning-Based Approach for Hardware Faults Prediction," *IEEE Transactions on Circuits and Systems I: Regular Papers*, vol. 67, no. 11, pp. 3880-3892, 2020.
[11] Deepak Sharma, Pravin Chandra, "Software Fault Prediction Using Machine-Learning Techniques," *Smart Computing and Informatics*, vol. 78, 29 October 2017.
[12] Aurick Qiao, Bryon Aragam, Bingjing Zhang, Eric P. Xing, "Fault Tolerance in Iterative-Convergent Machine Learning," *36th International Conference on Machine Learning, PMLR97*, pp. 5220-5230, 2019, Long Beach, CA, USA.
[13] Taiwo Oladipupo Ayodele, "Types of Machine Learning Algorithms," InTech, 2010.
[14] Ayon Dey, "Machine Learning Algorithms: A Review," *International Journal of Computer Science and Information Technologies*, vol. 7, no. 3, pp. 1174-1179, 2016.
[15] Gongde Guo, Hui Wang, David Bell, Yaxin Bi, Kieran Greer, "KNN Model-Based Approach in Classification," In: Meersman R., Tari Z., Schmidt D. C. (eds), On The Move to Meaningful Internet Systems

2003: CoopIS, DOA, and ODBASE. OTM 2003. Lecture Notes in Computer Science, vol. 2888. Springer, Berlin, Heidelberg.

[16] Ashis Pradhan, "Support Vector Machine – A Survey," *International Journal of Emerging Technology and Advanced Engineering*, vol. 2, no. 8, pp. 82-85, 2012.

[17] Achmad Widodo, Bo-Suk Yang, "Support Vector Machine in Machine Condition Monitoring and Fault Diagnosis," *Mechanical Systems and Signal Processing*, vol. 21, no. 6, pp. 2560-2574, 2007.

[18] Hitesh Mohapatra, Amiya Kumar Rath, "Fault-Tolerant Mechanism for Wireless Sensor Network," *IET Wireless Sensor Systems*, vol. 10, no. 1, pp. 23-30, 2020.

[19] Samira Chouikhi, In'es El Korbi, Yacine Ghamri-Doudane, Leila Azouz Saidane, "A Survey on Fault Tolerance in Small and Large Scale Wireless Sensor Networks," *Computer Communications*, vol. 69, pp. 22-37, 2015.

[20] Mohammad Abu, Alsheikh, Shaowei Lin, Dusit Niyato, Hwee-Pink Tan, "Machine Learning in Wireless Sensor Networks: Algorithms, Strategies, and Applications," *IEEE Communications Surveys & Tutorials*, vol. 16, no. 4, pp. 1996-2018, 2014.

[21] N. Satheesh, M. V. Rathnamma, G. Rajeshkumar, P. Vidya Sagar, Pankaj Dadheech, S. R. Dogiwal, Priya Velayutham, Sudhakar Sengan, "Flow-based Anomaly Intrusion Detection using Machine Learning Model with Software Defined Networking for OpenFlow Network," *Micro Processors and Microsystems*, vol. 79, 103285, 2020.

[22] Raouf Boutaba, Mohammad A. Salahuddin, Noura Limam, Sara Ayoubi, Nashid Shahriar Felipe Estrada-Solan, Oscar M. Caicedo, "A Comprehensive Survey on Machine Learning for Networking: Evolution, Applications, and Research Opportunities," *Journal of Internet Service and Application*, vol. 9, 16, 2018.

[23] Saad Ahmad Khan, Ladislau Bölöni, Damla Turgut, "Bridge Protection Algorithms Technique for Fault Tolerance in Sensor Networks," *Ad Hoc Networks*, vol. 24, part A, pp. 186-199, 2015.

[24] Martin Radetzki, Chaochao Feng, Xueqian Zhao, Axel Jantsch, "Methods for Fault Tolerance in Networks-on-Chip," *ACM Computer Survey*, vol. 46, no. 1, pp. 1-38, 2013.

[25] Shi Jin, Zhaobo Zhang, Krishnendu Chakrabarty, "Towards Predictive Fault Tolerance in a Core-Router System: Anomaly Detection Using Correlation-Based Time-Series Analysis," *IEEE Transactions on*

Computer-Aided Design of Integrated Circuits and Systems, vol. 37, no. 10, pp. 2111-2124, 2018.

[26] Samurdhi Karunaratne, Haris Gacanin, "An Overview of Machine Learning Approaches in Wireless Mesh Networks," *IEEE Communications Magazine*, vol. 57, no. 4, pp. 102-108, 2019.

[27] Mowei Wang, Yong Cui, Xin Wang, Shihan Xiao, Junchen Jiang, "Machine Learning for Networking: Workflow, Advances, and Opportunities," *IEEE Network*, vol. 32, no. 2, pp. 92-99, 2018.

[28] Yas A. Alsultanny, "Fault Tolerance Effect on Computer Networks Availability," IADIS International Conference e-Learning, 2013.

[29] Tsang-Yi Wang, Yunghsiang S. Han, Pramod K. Varshney, Po-Ning Chen, "Distributed Fault-Tolerant Classification in Wireless Sensor Networks," *IEEE Journal on Selected Areas in Communications*, vol. 23, no. 4, pp. 724-734, 2005.

[30] Yunxia Feng, Shaojie Tang, Guojun Dai, "Fault-Tolerant Data Aggregation Scheduling with Local Information in Wireless Sensor Networks," *Tsinghua Science and Technology*, vol. 16, no. 5, pp. 451-463, 2011.

[31] Myeong-Hyeon Lee, Yoon-Hwa Choi, "Fault Detection of Wireless Sensor Networks," *Computer Communications*, vol. 31, no. 14, pp. 3469-3475, 2008.

Index

A

Abnormal pattern 71
Adversarial learning 236
Adversarial multimedia forensics 236
Analysis tools viii, xx, 12, 177
Android viii, 87, 99, 122, 219
Android application dissection viii, 87, 94, 96
Android application package xv, 87, 91, 99
Android malware detection 87, 101, 102, 232
Android mobile applications 87
Artificial Intelligence xi, 130, 215, 230
Attack models 49
Advanced Persistent Threats 7

B

Big Data xiii, 166, 215, 231
Big Data Analytics xiii, xiv
Blockchain ix, 129, 140, 151, 161
Boosted Tree Learning 195

C

Clustering 71, 75, 80, 178
Convolutional neural networks 157, 193, 236, 238
Cyber threat intelligence 4, 20, 30, 32
Cyberattack attribution 4
Cybersecurity 4, 31, 130, 159
Cyberthreat x, 176, 181,

D

Data privacy ix, 142
DeepFake 236, 252
Distributed network 175, 254
Deep Neural Network 178, 192, 219, 232
Deep Learning 46, 72, 130, 178

E

Encrypted traffic 71, 72, 80, 85
Ensemble methods 196
Encryption 54, 71, 123, 150

F

Failure node detector 254
Fault tolerance xviii, 253, 262, 271, 273

G

Generative adversarial networks 235, 236, 251

H

Hybrid Network x, 175, 192, 194
Hybrid NIDS 176, 180, 183, 191

I

Information flow control 105
Internet of Things 25, 49, 139, 167
Intrusion detection 15, 36, 145, 175
Indicator of Compromise 22
Intrusion Detection System 145, 177, 193, 221
Intrusion Prevention xiii
Intelligent Systems xiv, 29, 168,

K

K-nearest neighbor 73, 146, 199, 260, 263

L

Location-based services 49, 50, 68
Logistic regression 106, 119, 211, 226

M

Machine Learning 46, 85, 105, 152
Malware, 92, 101, 162, 195
Malware detection xi, 87, 139, 208
Malware Analysis xiv, 89, 95, 101

N

Naïve Bayes xvii, 72, 106, 213

P

Privacy preservation 49, 67
Privacy Protection vii, 53, 56, 65
Privilege escalation 78, 102, 105, 126

R

Ransomware xvii, 134, 196, 214
Ransomware Detection xi, 213, 218, 223
Random Forest 71, 80, 121, 178

S

Sharing platforms v, 3, 8, 26
Smart contracts xx, 131, 142, 158
Software Defined Network 272, 273
Stacked ensembles xi, 196, 201
Supervised learning vii, 71, 75, 82
Support Vector Machine 146, 198, 256, 258

T

Threat intelligence v, xiv, 21, 26
Tree-based methods 196, 208
Threat Analytics 5

U

Unauthorized intent receipt 105, 107, 108
Unsupervised learning 71, 75, 82, 159

W

Web application firewall vi, 35, 36, 47
Web application security 35, 46, 47
Web attacks 36, 41, 133

About the Editors

Yassine Maleh (https://orcid.org/0000-0003-4704-5364) is currently an Associate Professor of cybersecurity and IT governance with Sultan Moulay Slimane University, Morocco. He is the founding chair of IEEE Consultant Network Morocco and the founding president of the African Research Center of Information Technology & Cybersecurity. He is a senior member of IEEE and a member of the International Association of Engineers IAENG and The Machine Intelligence Research Labs. Dr. Maleh has made contributions in the fields of information security and privacy, Internet of Things security, and wireless and constrained networks security. His research interests include information security and privacy, Internet of Things, networks security, information system, and IT governance. He has published over 100 papers (book chapters, international journals, and conferences/workshops), 20 edited books, and 3 authored books. He is the editor-in-chief for the *International Journal of Information Security and Privacy* and the *International Journal of Smart Security Technologies* (IJSST). He serves as an associate editor for *IEEE Access* (2019 Impact Factor 4.098), the *International Journal of Digital Crime and Forensics* (IJDCF), and the *International Journal of Information Security and Privacy* (IJISP). He is a series editor of *Advances in Cybersecurity Management*, by CRC Taylor & Francis. He was also a guest editor of a special issue on *Recent Advances on Cyber Security and Privacy for Cloud-of-Things* of the *International Journal of Digital Crime and Forensics* (IJDCF), Volume 10, Issue 3, July–September 2019. He has served and continues to serve on executive and technical program committees and as a reviewer of numerous international conferences and journals such as Elsevier Ad Hoc Networks, IEEE Network Magazine, IEEE Sensor Journal, ICT Express, and Springer Cluster Computing. He was the publicity chair of BCCA 2019 and the general chair of the MLBDACP 19 Symposium and ICI2C'21 Conference. He received Publon Top 1% reviewer award for the years 2018 and 2019.

Mamoun Alazab (https://orcid.org/0000-0002-1928-3704) is currently an Associate Professor with the College of Engineering, IT and Environment, Charles Darwin University, Australia. He received the Ph.D. degree in computer science from the Federation University of Australia, School of Science, Information Technology and Engineering. He is a cybersecurity researcher and practitioner with industry and academic experience. Dr Alazab's research is multidisciplinary that focuses on cybersecurity and digital forensics of computer systems including current and emerging issues in the cyber environment like cyber-physical systems and the Internet of Things, by taking into consideration the unique challenges present in these environments, with a focus on cybercrime detection and prevention. He looks into the intersection use of machine learning as an essential tool for cybersecurity, for example, for detecting attacks, analyzing malicious code, or uncovering vulnerabilities in software. He has more than 100 research papers. He is the recipient of short fellowship from Japan Society for the Promotion of Science (JSPS) based on his nomination from the Australian Academy of Science. He delivered many invited and keynote speeches, 27 events in 2019 alone. He convened and chaired more than 50 conferences and workshops. He is the founding chair of the IEEE Northern Territory Subsection (February 2019 to present). He is a senior member of the IEEE, Cybersecurity Academic Ambassador for Oman's Information Technology Authority (ITA), member of the IEEE Computer Society's Technical Committee on Security and Privacy (TCSP), and has worked closely with government and industry on many projects, including IBM, Trend Micro, the Australian Federal Police (AFP), the Australian Communications and Media Authority (ACMA), Westpac, UNODC, and the Attorney General's Department.

Loai Tawalbeh (https://orcid.org/0000-0002-2294-9829) received the Ph.D. degree in electrical & computer engineering from Oregon State University in 2004, and the M.Sc. in 2002 from the same university with GPA 4/4. Dr. Tawalbeh is currently an Associate Professor with the Department of Computing and Cyber Security, Texas A&M University-San Antonio. Before that, he was a Visiting Researcher with the University of California-Santa Barbra. Since 2005, he has been teaching/developing more than 25 courses in different computer engineering disciplines and science with a focus on cybersecurity for the undergraduate/graduate programs at the NewYork Institute of Technology (NYIT), DePaul's University, and Jordan University of Science and Technology. Dr. Tawalbeh won many research grants and awards with over 2 Million USD. He has over 80 research publications in refereed international journals and conferences.

Imed Romdhani has been an Associate Professor in networking with Edinburgh Napier University since June 2005. He received the Ph.D. degree from the University of Technology of Compiegne (UTC), France, in May 2005. He also holds an engineering and a Master's degree in networking obtained, respectively, in 1998 and 2001 from the National School of Computing (ENSI, Tunisia) and Louis Pasteur University of Strasbourg (ULP, France). He worked extensively with Motorola Research Labs in Paris and authored four patents.